Foresight for Dynamic Organisations in Unstable Environments

Foresight has an especially important role to play in times of trouble and transition. When turbulence and radical uncertainty are the framework in which organisations operate, it arises as a tool for intellectual freedom and enhanced strategic leverage. But foresight itself is always in a continuing process of innovation, as it is not detached from the changing environment that engulfs decision-makers and stakeholders. Taking stock of recent developments in foresight methodologies and implementation experiences has now emerged as an urgent task as new experiments accumulate in the twenty-first century. Looking ahead calls for both a review of contemporary perspectives and innovative practices in foresight methodology, and for salient experiments on how foresight is embedded in organisations.

This book brings together real-world cases and conceptual proposals to bridge the gap between practitioners and researchers in the field of futures research. Such an ambition is an increasingly difficult balancing act as the gap between the needs of organisational leaders and the incentives of academics becomes an ever widening gulf due to increasingly specialised and self-absorbed agendas. An international team of authors from a wide variety of backgrounds come together in this book to provide first-hand views from the frontier of empirical work and theoretical reflection on foresight.

This book was originally published as a special issue of *Technology Analysis & Strategic Management.*

Sandro Mendonça is a lecturer at the Lisbon University Institute (ISCTE-IUL), Lisbon, Portugal. He is a researcher at UNIDE and Dinamia (research units of ISCTE-IUL) and UECE (research unit of ISEG-UTL). He has pursued a research agenda in the fields of the economics of technical change and strategic foresight, but has developed a growing interest in the areas of economic history and peace studies. His work has been published in a range of journals including *Research Policy, Industrial and Corporate Change, Long Range Planning, Technology Analysis and Strategic Change, Technological Forecasting and Social Change, Futures* and *Journal of Economic Issues.* He is a member of the executive board of Obercom (the Portuguese media watch) and member of the executive committee of EconTel (Telecommunications Economics COST Network). In the capacity of expert he has cooperated with the European Science Foundation, the European Commission and the European Committee for Standardization (CEN). He is a consultant on matters of corporate innovation strategy and intellectual property public policy.

Bartolomeo Sapio received his Doctor Laurea Degree in Electronic Engineering summa cum laude at the University of Rome 'La Sapienza', and is currently a researcher with Fondazione Ugo Bordoni. He has carried out methodological research in the field of scenario analysis, developing the original methods WISE, SEARCH and GIMMICKS, and applying them to multimedia, broadband networks, the Internet, mobile services and the convergence between fixed and mobile networks. He researched the diffusion of digital television and adoption patterns of interactive services. He has participated in several International projects and is currently chairman of COST (COoperation in the field of Scientific and Technical research) Action 298 "Participation in the Broadband Society".

Foresight for Dynamic Organisations in Unstable Environments

A Search for New Frameworks

Edited by

Sandro Mendonça and Bartolomeo Sapio

Routledge
Taylor & Francis Group
LONDON AND NEW YORK

First published 2011
by Routledge
2 Park Square, Milton Park, Abingdon, Oxon, OX14 4RN

Simultaneously published in the USA and Canada
by Routledge
711 Third Avenue, New York, NY 10017

Routledge is an imprint of the Taylor & Francis Group, an informa business

This book is a reproduction of *Technology Analysis & Strategic Management*, volume 21, issue 3. The Publisher requests to those authors who may be citing this book to state, also, the bibliographical details of the special issue on which the book was based.

British Library Cataloguing in Publication Data
A catalogue record for this book is available from the British Library

ISBN 13: 978-0-415-67223-8

Typeset in Times New Roman
by Taylor & Francis Books
Printed and bound by TJI Digital, Padstow, Cornwall

Disclaimer
The publisher would like to make readers aware that the chapters in this book are referred to as articles as they had been in the special issue. The publisher accepts responsibility for any inconsistencies that may have arisen in the course of preparing this volume for print.

Contents

Notes on contributors

Toni Ahlqvist is a Senior Research Scientist at VTT Technical Research Centre of Finland and a postdoctoral researcher at the Academy of Finland. His research is focused on technology foresight, technological transformation of societies, innovation systems and economic geography. He holds a PhD in human geography from the University of Turku, Finland.

Erik den Hartigh is an assistant professor of Strategy and Innovation at the School of Economics and Administrative Sciences of Özyeğin University (Istanbul, Turkey). He is also affiliated as an assistant professor with the Department of Technology, Strategy and Entrepreneurship of the Delft University of Technology. His research is about corporate strategy, innovation and business networks. He teaches courses in general management, corporate strategy, technology strategy, innovation strategy and innovation systems.

Annele Eerola is Senior Research Scientist, of VTT Knowledge Centre 'Organisations, Networks and Innovation Systems'. Her research focuses on the links between foresight knowledge, corporate strategy and innovation policy. She holds a PhD from Helsinki Swedish School of Economics and Business Administration, and Tech. Lic. degree from Helsinki University of Technology.

Dr Olivier Glassey is a sociologist who studies online mass collaboration and emergent collective behaviours in several domains (virtual communities, social networks, Wikipedia, etc.). He is leading the research unit Social Studies of Science and Technology in the Observation of Science, Policy and Society at the University of Lausanne.

Attila Havas (MA 1983, PhD 1997) is a Senior Research Fellow at the Institute of Economics, Hungarian Academy of Sciences. His academic interests are in economics of innovation, theory and practice of innovation policy, and technology foresight. In 1997–2000 he was Programme Director of TEP, the Hungarian Technology Foresight Programme. He has participated in international research projects on STI policies, innovation and transition, as well as on foresight and prospective analyses, and been a member of several EU expert groups. He has advised national governments and international organisations on the above issues.

Ulrik Jørgensen is a MSc in Electronics Engineering and has a PhD in Innovation Economics. He is professor (docent) of technology assessment at the Department of Management Engineering at the Technical University of Denmark. He is head of the Danish research school for Design and Innovation and leads the strategic research project 'Temporary spaces for user driven innovation'. His research is in the field of science and technology studies with emphasis on design processes, engineering competences, and sustainable transitions.

Sirkku Kivisaari works as Senior Research Scientist in VTT. Her research relates to future-oriented technology assessment and innovation studies. Her special interest lies in enhancing innovations provoked by societal concerns for wellbeing of the ageing society and for cleaner environment. She holds a PhD in business management from Helsinki School of Management.

Raija Koivisto is a Research Professor and team leader in VTT Knowledge Centre 'Organisations, Networks and Innovation Systems'. She has 25 years experience with VTT in safety, security, risk assessment and foresight related research and consultancy. She holds a docent position in the University of Oulu and lectures also in Lappeenranta University of Technology.

Totti Könnölä is Scientific Officer at the Institute for Prospective Technological Studies of the Joint Research Centre in the European Commission. He is also Adjunct Professor of Operations and Technology Management in the IE Business School. His work on innovation and environmental management and policy has been published widely.

Sandro Mendonça is a lecturer at the Department of Economics, ISCTE Lisbon University Institute. His work has been published in a wide range of journals including *Research Policy, Industrial and Corporate Change, Long Range Planning, Technology Analysis and Strategic Change, Technological Forecasting and Social Change, Futures* and *Journal of Economic Issues.* He is member of the executive board of Obercom (the Portuguese media watch) and member of the executive committee of EconTel (Telecommunications Economics COST Network). He is a consultant on matters of corporate innovation strategy and intellectual property public policy.

Gianluca Misuraca is currently a Scientific Officer at the Institute for Prospective Technological Studies of the European Commission's Joint Research Centre (JRC-IPTS) in Seville, Spain, where he is part of the Information Society Unit focusing on research in the area of ICT for Governance and Policy Modelling. Gianluca is also a Research Associate at the Ecole Polytechnique Fédérale de Lausanne (EPFL) where, before joining IPTS, he was the Managing Director and Scientific Coordinator of the Global Executive Master in e-Governance (eGov), organised by EPFL in collaboration with prestigious institutions worldwide. Previously, he worked for the United Nations Department of Economic and Social Affairs (UNDESA) acting as Liaison Officer for Africa in knowledge management and e-Governance, being responsible for the design and of the e-Africa initiative for good governance and the implementation of the United Nations Online Network in Public Administration and Finance (UNPAN). During his career, Gianluca held several positions as policy advisor for different International Organisations and bilateral cooperation agencies as well as working with various consulting and industrial organisations in the area of e-Government, public administration reform, regional development, research, innovation and technological transfer.

Pierre Rossel holds a PhD in anthropology and is currently senior scientific fellow at the College of Management of Technology at the Ecole polytechnique fédérale of Lausanne (EPFL), Switzerland. His main research expertise is the technology assessment and foresight of information and communication technologies, covering a variety of issues and applications from cutting-edge telecom innovation to next generation internet services and digital cities, on which he has published extensively (books, chapters, papers, reports). Within the college of management at EPFL, he gives yearly one-semester courses, at Master level on 'Early detection of technological innovation potential', and at PhD level on "Technology foresight". He has recently initiated a new INTERREG project with the objective of boosting SME innovation capabilities in the French-Swiss Jura Arc micro-technology cluster.

Bartolomeo Sapio received his Doctor Laurea Degree in Electronic Engineering summa cum laude at the University of Rome 'La Sapienza', and is currently a researcher with Fondazione Ugo Bordoni. He has carried out methodological research in the field of scenario analysis, developing the original methods WISE, SEARCH and GIMMICKS, and applying them to multimedia, broadband networks, the Internet, mobile services and the convergence between fixed and mobile networks. He researched the diffusion of digital television and adoption patterns of interactive services. He has participated in several International projects and is currently Chairman of COST (COoperation in the field of Scientific and Technical research) Action 298 "Participation in the Broadband Society".

Jan Oliver Schwarz is a strategy consultant at an insurance company, working on strategy and strategic foresight. His research focuses on strategic foresight, scenario planning, trend research, and business wargaming. He holds a MA in General Management, an MPhil in Futures Studies and a PhD on strategic foresight from the Berlin University of the Arts, Germany.

Michael Søgaard Jørgensen is a MSc in Chemical Engineering and has a PhD in Technology Assessment. He is associate professor in user participation in technology assessment and technology development at Department of Management Engineering at Technical University of Denmark. His research areas are environmental management in companies and product chains, sustainable transition strategies and experiences, and technology foresight as part of innovation studies and environmental assessments.

Sébastien Treyer holds a background in environmental engineering, a PhD in environment management, and specialises in foresights for environmental policies. He has been in charge of developing foresights at the French ministry for environment, worked as a researcher at AgroParisTech, and is now director of programmes at Iddri (institute for sustainable development and international relations, Paris), a think tank specialised in policy research and negotiations for environmental issues necessitating coordination at the international scale. He has recently been involved in the "Agrimonde" project, producing scenarios about how to feed the world in 2050 in a sustainable way, and also worked on foresights for environmental policies at the local (national parks) or regional level (water basin authorities).

Patrick A. van der Duin is an assistant professor at Delft University of Technology, section Technology, Strategy and Entrepreneurship. He received his Masters degree in macro-economics at the University of Amsterdam, after which he worked as an applied scientist for six years at KPN Research. He received his PhD at Delft University of Technology. His main focus is on futures research, innovation management and systems, and how futures research and innovation are related to each other. He is also member of the editorial boards of Futures, Foresight, and the International Journal of Innovation and Technology Management.

Introduction: new perspectives and practices for managing foresight in changing organisational settings

Foresight in times of transition and trouble

In times characterised by a tyranny of turbulence foresight arises as a tool for intellectual freedom and enhanced strategic leverage, but, it should be noted, foresight itself is always in a continuing state of experimentation as it is not detached from the changing environment that engulfs organisations. Taking stock from developments in foresight methodologies and implementation experiences is especially important at this moment as the first decade of the present century draws to a close. Looking ahead calls for a review of new perspectives and recent practice on foresight methodology and on how foresight is embedded in organisations.

It might be argued that foresight is needed the most in circumstances in which uncertainty is the most severe and we believe that these are times in which foresight experts and practitioners should rise to the call. The few years into the twentieth-first century have, so far, only heightened a profound sense of immersion into the unthinkable. Citizens, communities and governments have been caught between the shock of dramatic discontinuities and the pressure of powerful long-term trends.

On the one hand, the stability of existing socio-economic structures has been violently pierced by events such as the epoch-defining terrorist attacks on American soil, the onslaught of major natural accidents like the Indian Ocean tsunami and hurricane Katrina, or a financial crisis in Wall Street which morphed into a sharp global economic downturn (variously described as an 'economic Pearl Harbour' and 'once-in-a century credit tsunami' by personalities such as legendary investor Warren Buffett and former US Federal Reserve chairman Alan Greenspan, respectively). On the other hand, the gradual development of trajectories in areas such as climate change, trade globalisation, income inequality, scientific acceleration, technological transition into a networked knowledge economy and the redistribution of geo-political protagonism to emerging powers such as China, Russia and India. Such trends have, slowly but steadily, bitten into the sustainability of old assumptions and perceived options in fields ranging from energy security, through food policy, to peace-building.

The exposure to uncontrollable change taking place — but at different speeds contemporaneously — in a great diversity of dimensions, and met with different levels of surprise by different actors and stakeholders, creates the potential for alarmist responses that undercut each other at several levels leaving no single agenda contradiction-free. Quick twists and turns in trends invite decision-makers to respond in adaptive, improvised ways, but at the risk of intertemporal inconsistency. In highly volatile systems moments presenting wholly unique threats and extraordinary opportunities succeed each other at a higher speed than individuals can assimilate, let alone filter and assess them. Constant change may be disarming and demobilising, but it also brings about an excess of half-baked advice. The persistence of turmoil also puts a premium on order; hence, the perpetually renewed proclamation of new 'game-changing' ideas in airport bookshop shelves and the recourse to history as treasure-chest of anecdotes and hints for the future in newspaper columns. Insurance against turbulence, and its traps, is on demand.

In such an environment shaken by incessant, fast-paced change foresight provides a sort of epistemic ballast to organisations. Foresight cushions attention-strapped organisations when in critical times short-run decisions have a critical bearing on long-term performance. Foresight can mediate the everyday interaction of social groups with topics that have a strategic bearing on the future of the economy and the ecosystem, thereby shaping perceptions and practices. In this way foresight can help to integrate heterogeneous and competing agendas, and introduce some degree of dynamic coherence into debates over policy and strategy.

We take foresight to be a deliberate, critical, reflexive and creative forward-looking engagement with future, action-dependent states of affairs. It distinguishes itself from utopia by the consideration of multiple future possibilities and the weight given to possible dystopian outcomes of evolution; it distinguishes itself from fortune-telling by the use of explicit methodologies open to criticism and improvement; it distinguishes itself from forecasting by admitting qualitative techniques and by not giving a misleading appearance of neutrality regarding the qualities (i.e. desirability) of future outcomes. Foresight is about relentless assumption questioning and about the transformation of the parameters that make us understand the future. It is mostly about discovering new options, not so much predicting by extrapolating past routines. Foresight can thus be defined as a research-based activity that draws on a wide array of disciplines and insights in order to build anticipatory and actionable knowledge useful for stakeholders and decision-makers.

Private and public organisations are still a prominent mode of coordinating collective human action in spite of the growth of peer-to-peer, digitally-empowered and mass distributed work in open social and business webs. From this perspective it is of particular interest, both theoretical and practical, to understand the new and evolving ways in which the systematic exploration of the future is being organised. Organisations are situated in dynamic and unpredictable environments, so they have to be innovative and resourceful. The new ways in which foresight is becoming embedded and managed in organisations constitutes the prime focus of this special issue.

A portfolio of foresight perspectives and practices

In today's globalised and technology-rich economy pursuing long-run sustainable routes that are robust enough to withstand short-run disruptions is a restless process that continuously mixes reflection and action. To further this special issue's goals we had the opportunity to assemble an international team of authors coming from a variety of backgrounds. This special issue brings together a sample of real-world experiments and of conceptual proposals designed to bridge between practitioners and researchers in the field of foresight. Such an ambition is an increasingly difficult balancing act as the gap between the needs of organisational leaders and the incentives of academics becomes an ever widening gulf as a result of increasingly specialised and self-absorbed agendas.

Jan Schwartz presents 'business wargaming', a simulation technique that can be implemented to make several teams within a given company more aware of their own interacting dynamics, of their competitors and the market. Lessons learned in this safe environment are of value to different organisational stakeholders, including top managers, workers and mid-level team leaders, who have to become more and more multifunctional to deal with complex business missions in which tolerance for error is declining. In his pioneering contribution Schwartz articulates how 'business wargaming' can become a tool in the arsenal of strategic foresight.

Pierre Rossel's paper refers to the role of weak signal analysis as a broad canvas in which organisational stakeholders can appraise and re-appraise strategising hypotheses in rugged decision-making environments. Rossel deals with the issues of examination of potential seeds of future change by directly confronting the question that pattern-recognition and pattern-projection may be indistinguishable in real time, that is, that observing reality and constructing it overlap. To capitalise on the paradox Rossel tackles a number of metaphors in the literature and advances the debate by introducing the twin notions of flexible framing and meta-framing as a conceptual matrix in which pattern identification efforts can take place in analytically reflexive conditions.

Sébastien Treyer's reflection starts out by a clear and simple, but often dismissed, premise; that foresight projects are typically initiated in already existing strategic conversations. Foresight projects are no mere one-off operations; they are exercises that play at the same time with old and new elements. An implication is that the effectiveness of foresight capabilities within organisations is partly a function of the success in both re-parameterising the representations of the future while keeping a link to past references that allow a community of participants to accumulate insights and knowledge. This conceptual framework allows us to stretch current best practice in foresight by providing a ground on which to erect criteria for quality evaluation of foresight interventions, to use the author's apt terminology.

The paper by Patrick van der Duin and Eric den Hartigh examines the tension between the technological innovation process and the strategic formulation process. Their argument is that foresight serves as a weight/counter-weight system between processes that are designed to open up options (innovation) and close down alternatives (strategy). The authors articulate their view on how foresight research interplays with innovation management and strategy making on the basis of a survey of multi-national companies. In their words, foresight as an integrative tool in the 'battle for the future' has different uses in different theatres of operations. A framework that combines stability vs dynamism of the internal and external environment of business firms is used to qualify the role of foresight in a setting of four ideal-type situations.

Michael Søgaard Jørgensen and Ulrik Jørgensen tell the story of a 'green' foresight project initiated by a Danish public organisation. In doing so their paper sheds light into how prospective environmental opportunities and threats can be managed in a way that takes into account the necessary contestability of possible social and technical solutions that may emerge as priorities. The discursive approach that Jørgensen and Jørgensen describe constitutes a model of how the complex relationship between social visions and technological options may be at the same time problematised and, simultaneously, held out as an opener of innovative policy pathways that were not predicted by stakeholders.

Gianluca Misuraca's paper charts the course from the reality of e-Government of first generation ('e-Gov 1.0') to the rhetoric of 'e-Gov 2.0'. On this dynamic background the author discusses the future promises of m-Government. The paper further addresses the various emerging scenarios and how they can be used in a policy relevant way, suggesting that m-Government appears to be a promising area for research and experimentation, opening new possibilities for delocalised governmental servicing and user involvement.

The team of authors headed by Totti Könnölä offers an analysis of what may be termed 'modular' foresight projects at a leading contract research organisation, the VTT, Technical Research Centre of Finland. What marks the VTT case out is the huge variety of foresight projects managed and the variegated range of foresight expertise involved. This team of authors offers an account of how it is possible to make sense of a widening array of exercises by developing a framework for the integrated management of the different kinds of foresight activities. Modular process design arises as a sensible option to minimise the cost of running such activities while taking advantage of the cross-fertilisation potential of available expertise.

Attila Havas focuses on a long-established institution, the University, and discusses how novel methods are needed to shape the future of the higher education system. Havas suggests multi-level 'cascading' futures framework as key trends and driving forces are international in their nature and universities are embedded in broader stakeholder systems. What is proposed is multi-level foresight as a way to appraise desired and feasible future states and, more broadly, as a source of higher education renewal.

This special issue grew out of the final conference of COST A22 network entitled 'From Oracles to Dialogue: Exploring New Ways to Explore the Future', held at the National Technical University, Athens, Greece, 9–11 July 2007.

Acknowledgements

A number of researchers and practitioners contributed to the collective effort of reviewing the papers. We greatly appreciate the contribution of Alexandre Abreu, Anders Koed Madsen, Andrea Pannone, Andreas Graefe, António Ioris, Barend van der Meulen, Christof Weinhardt, Cristiano Cagnin, Eugene Loos, Jan Schwartz, Jon Mikel Zabala Iturriagagoitia, Jonathan Snapsed, Jorge Vieira, José Luís de Almeida e Silva, Magnus Gulbrandsen, Miguel Lopes, Miguel Pina e Cunha, Olivier da Costa, Patrícia Palma, Patrick Law, Pedro Puga, Peter Hall, Pierre Rossel, Roberto Poli, Serge Stalpers, Simone Arnaldi, Tiago Mata, Tomaz Turk and Totti Könnölä. This special issue also owes to the encouragement and editorial support of Harry Rothman and Sean Rothman.

Business wargaming: developing foresight within a strategic simulation

Jan Oliver Schwarz

Berlin University of the Arts, Germany

The development of foresight is a crucial activity for any organisation, especially in times of increasing dynamism and complexity. The aim of this article is to discuss the usefulness of developing foresight in a business wargame. Wargaming, a role-playing simulation of a dynamic situation, has been used mainly by the military, but more recently, it has also been used in the context of competitive intelligence and strategy-testing in organisations. A business wargame typically evolves several years into the future, allowing the participants to 'experience' future dynamics in the organisational environment and, consequently, to develop foresight. This article argues that, because of its participative and dynamic nature, business wargaming can overcome cognitive barriers, challenge mental models, detect weak signals of change in an organisational environment, re-direct attention in an organisation, and assist an organisation in developing foresight.

1. Introduction

Organisations are confronted with a more complex and dynamic environment than ever, one that is characterised by discontinuities and an uncertain future. This condition is likely not only to continue but to intensify. The major tasks for today's managers are to make decisions, formulate strategies, and implement strategic management systems; however, in such an environment, the imperative of 'predict and prepare', the foundation of the neoclassical school of management (Gharajedaghi 1999), is no longer appropriate. Corporations are now being urged to develop foresight (Hamel and Prahald 1994; Courtney 2001), to prepare for the future, not just to predict it. In addition, several governments have already engaged in national foresight activities, a trend that is also likely to accelerate (Martin and Johnson 1999).

In practise, the dominant question for organisations remains, how to develop and implement foresight processes. While the scenario technique has been praised as valuable in developing foresight, numerous barriers, especially cognitive ones, exist (De Geus 1997; Bazerman and Watkins 2004; Day and Schoemaker 2004; MacKay and McKiernan 2004; Seidl and Van Aaken 2004); these cognitive barriers tend to hinder foresight (Schwarz 2005).

The primary aim of this article is to examine how business wargaming could contribute to developing foresight and to reducing the organisational barriers to developing foresight. Business wargaming has been applied for several centuries to the military field and since the late 1980s management has used it for instance to test strategies. In short, a business wargame is a role-playing simulation of a dynamic business situation (Kurtz 2003).

The article is organised as follows: first, the role and the importance of developing foresight will be discussed; second, the scenario technique will be critiqued and the cognitive barriers to developing foresight will be discussed; third, the method of business wargaming will be introduced; and fourth, the advantages of business wargaming and its role in the foresight process will be explored.

2. Developing foresight

As early as 1931, Alfred North Whitehead, in a celebrated lecture at the Harvard Business School, identified foresight as the crucial feature of the competent business mind (Tsoukas 2004). Since the late 1980s, foresight has been used to describe activities that inform decision-makers, by providing input concerning the long-term future (Miles, Keenan, and Kaivo-Oja 2003). However, the term is often used interchangeably with terms such as futures studies and prospective analysis, so there is little consensus on the boundaries between these fields.

Makridakis (2004, XIII) states that the role of foresight is 'to provide business executives and government policy makers with ways of seeing the future with different eyes and fully understanding the possible implications of alternative technological/societal paths'. Foresight involves spotting developments before they become trends, seeing patterns before they fully emerge, and grasping the relevant features of social currents that are likely to have an impact (Tsoukas 2004), rather than making predictions. Ideally, foresight recognises even faint signals of change or trends in the corporate business environment and imagines alternative pictures and how organisations will evolve as a part of these pictures.

As a process in an organisational context, foresight is based on the principle that discontinuities (in most cases) do not emerge without warning, and these warning signs can be described as 'weak signals' (Ansoff 1975). Foresight aims at early detection of these signals, which could lead to strategic surprises and to an event with the potential to jeopardise the organisation's strategy. Day and Schoemaker (2006, 1–2) articulate the purpose of spotting weak signals as providing the opportunity to 'quickly spot those signals that are relevant and explore them further, filter out the noise, and pursue opportunities ahead of the competition or recognise the early signs of trouble before they escalate into major problems'. The role of a foresight process in an organisation is not so much to predict the future, but to prepare for it (Tsoukas and Shepherd 2004).

Chia (2004) argues that foresight is achieved in an organisation by re-educating management. There are several ways for an organisation to generate foresight. The scenario technique (Fahey and Randall 1998) or Delphi iterations (Helmer 1983) can be applied to develop foresight, but a full foresight process, which incorporates these tools and other processes, is more promising. The argument has been made that a foresight process (Horton 1999) involves three phases, this has in particular been recognised in the German discussion on foresight processes, refereed to as *Strategische Frühaufklärung* (Müller 1981; Krystek and Müller-Stewens 1993; Liebl 1996, 2000):

(1) Information gathering, achieved by scanning the organisational environment (Aguilar 1967) for weak signals, which can also be perceived as trends.

(2) Diagnosing and interpreting the detected trends.
(3) Formulating reaction strategies in response to those trends.

3. Scenario criticisms and cognitive barriers

Scenarios or scenario planning are crucial to developing foresight, but scenario planning has been the subject of criticism. Understanding this criticism, in addition to the cognitive barriers to the development of foresight, will add texture to the discussion on business wargaming.

3.1. *Scenario planning and its criticism*

Scenario planning has its roots in the scenario technique, which was developed at the Rand Corporation in the 1950s (Pohl 1996). In the 1970s, it was elaborated and refined by Royal Dutch/Shell, which integrated the scenario technique with strategic planning (Fahey and Randall 1998). The aim of scenarios is to feature trends, alternatives, key uncertainties, and to combine them into pictures of the future, not covering all eventualities but defining the substance and the boundaries of future outcomes (Schoemaker 1992). In addition to the important task of providing alternative pictures of the future, scenarios and scenario planning can restore complexity to the strategic planning process and challenge the mental models of the managers involved.

However, scenario planning is vulnerable to the charge that scenarios struggle with incorporating future dynamics of markets. Oriesek and Friedrich (2003) pointed out that the scenarios used in scenario planning processes are often little more than projections crafted onto past experiences. They criticise scenario planning and the standard planning process for relating less to the future than to the past and, therefore, allowing the organisation to be ambushed by unforeseen events. MacKay and McKiernan's (2004) criticism claims that scenarios often fail to spot the weak signals of change in an organisational environment, an argument which also has been reiterated by others (Liebl 2004; MacKay and McKiernan 2004; Neugarten 2006; Postma and Liebl 2005). These authors contended that the knowledge basis of a scenario exercise is often unclear, which suggests that mental models have not been sufficiently challenged. Kurtz (2003) comments that criticism of both planning and scenario planning concepts have evolved from the failure to enlist the participation of enough management staff in a learning exercise to confront, understand and deal with the dynamics of particular business situations. Schwarz (2006, 2008) complains that scenario exercises tend to be time-consuming and complicated. These criticisms of scenario planning reveal some room for improvement for approaches to the development of foresight in an organisation. However, the manner in which a scenario exercise is carried out is just as critical.

3.2. *Cognitive barriers to foresight*

In addition to numerous operational reasons (Schwarz 2005), which hinder foresight, several authors have blamed cognitive barriers for the failure to develop foresight (De Geus 1997; Bazerman and Watkins 2004; Day and Schoemaker 2004; MacKay and McKiernan 2004; Seidl and Van Aaken 2004). Organisations often fail to perceive weak signals of change or trends because they do not fit the mental models of the organisation and are therefore rejected. Senge (1990, 8) defines mental models as 'deeply ingrained assumptions, generalisations, or even pictures or images that influence how we understand the world and how we take action'.

Day and Schoemaker (2006, 37) for instance describe obstacles to developing foresight as the 'powerful tendency to ignore warning signals and pretend that all is well. The more intelligent

Mental filters: Research shows that people tend to force the world into their existing frames. Weak signals that don't fit are typically distorted or ignored. Humans see what they expect to see, rather than what is there.
Overconfidence: A demonstrated tendency to be too certain also makes people tend to believe that the current view they hold is correct.
Penchant for confirming as opposed to disconfirming evidence: It is more difficult to detect disconfirming evidence than confirming evidence, so the mind is more likely to accept than to reject an idea.
Dislike for ambiguity: People dislike ambiguity, particularly in organizations in which managers are expected to have answers to questions.
Groupthink: Members of organizations take comfort from belonging to the majority and seeing the world in the same way, so there is a tendency to go along with what others say, rather than to use an individual mind to find flaw in the group's thinking.

Figure 1. Cognitive barriers (Day and Schoemaker 2004, 138).

we are, the better we also are at rationalising away important signals of impending doom.' The relevance of cognitive barriers in foresight is also underscored by the notion that cognitive activities are based on concepts of the future (Ingvar 1985) and that humans can only imagine the future in ways which are related to their experience and understanding of the past (Reading 2004).

Understanding how cognitive factors can interfere with the process is an important part of implementing a foresight process in an organisation. As depicted in Figure 1, these factors include mental filters. Ilmola and Kuusi (2006) have linked the hindering of foresight to filters of weak signals such as overconfidence, a penchant for confirming, rather than disconfirming evidence, dislike for ambiguity and groupthink.

4. Introducing business wargaming

The following sections describe the method of business wargaming and how business wargaming can be a vehicle for developing foresight; it also responds to criticism of scenario planning and cognitive barriers to foresight.

Kurtz (2003) describes a business wargame as a role-playing simulation of a dynamic business situation. Each team in the wargame is assigned to play a stakeholder, such as a competitor. The typical business wargame lasts several rounds, each one representing a predetermined time period. A business wargame is usually prefaced by extensive research on the industry in which the wargame is supposed to take place.

'Wargame' is the literal translation of the German *Kriegsspiel*. Since, because of the gravity of war, many people in the military are uncomfortable with the term *game*, wargames have been called 'map manoeuvres', 'field manoeuvres', 'exercises', or more recently, 'modelling and simulation'. The business world has its own reservations with the use of the terms 'war' and 'game', so wargames have also been referred to as 'strategic simulations' (Oriesek and Schwarz 2008).

Wargames can have several purposes, among which are strategy-testing, crisis planning and management, change management, planning, and training and education. Wargames have been applied in both the public and private sectors.

4.1. *History of business wargaming – military*

Business wargaming can be traced as far back as ancient Greece. It grew out of military wargaming, which was used to prepare generals and officers for unforeseen circumstances on the battlefield. Games about warfare have probably existed as long as warfare itself (Perla 1990). Perla (1990) credits the Chinese general and military philosopher Sun Tzu for inventing the first wargame, Wei-Hai', meaning 'encirclement', about 5000 years ago. 'Go', appeared around 2200 BCE, 'Chaturanga' around AD 500, and chess probably in the sixth century. All could be perceived as the successors of this wargame. Even though chess is much more abstract than other wargames, chess and its forerunners contain elements of warfare.

The proliferation of military wargaming increased in the nineteenth century when armies expanded, making them more difficult to move and industrialisation brought in artillery that was capable of reaching farther with faster repetition and greater precision. By the end of the nineteenth century, automatic machine guns had been introduced, increasing individual firepower, and the railroad made it possible to move troops faster. Wargames were used in Prussia, the USA, United Kingdom, Italy, France, Russia and Japan, especially after the Prussian victory in the Franco-Prussian war of 1870–1871. The USA, which continued to use wargaming after the Second World War, is regarded as the pacesetter in twentieth-century military wargaming. By the end of the twentieth century, the US military was using wargaming to enable its military strategies to impose some certainty on an uncertain future (Haffa and Patton 1999).

4.2. *History of business wargaming – management*

In 1957, the American Management Association (AMA) developed the first widely known business game, 'The AMA Top Management Decision Simulation' (Kalman and Rhenman 1975). The AMA game required five teams of three to five players each to make business decisions. The simulated company produced a single product, which was sold in an open and competitive market and the players represented company officers. Five to ten years of company operation were usually simulated in each game. In order to reduce the participants' computational burden, the AMA game allowed each company only a few decision alternatives and a mathematical model, aided by computers, was used to evaluate the teams' performance.

In 1958, the *Harvard Business Review* (Andlinger 1958) published an article on the application of wargaming to the business environment; the article referred to 'business gaming' and 'management simulation'. The games described in the article were primarily used for training and education and built on the military use of wargaming. These business games can be classified as general games, where top management makes the decisions that need to be made, or functional games, which concentrate on a company's functions, such as production, finance, or marketing.

Even though business wargaming had been applied to company planning in the 1950s (Bellman et al. 1957), wargaming was not widely and strategically adapted for business purposes until the mid 1980s (Ginter and Rucks 1984; Treat, Thibault, and Asin 1996). In the field of competitive intelligence, business wargames have drawn special attention for their focus on the competitors. More recently, business wargaming has emerged as a tool for competitive intelligence and for strategy formulation (Kurtz 2003).

4.3. *How does a business wargame work?*

A business wargame (Oriesek and Schwarz 2008) typically evolves over three moves and represents a certain length of time (e.g. 3–10 years). Four types of teams usually compete in

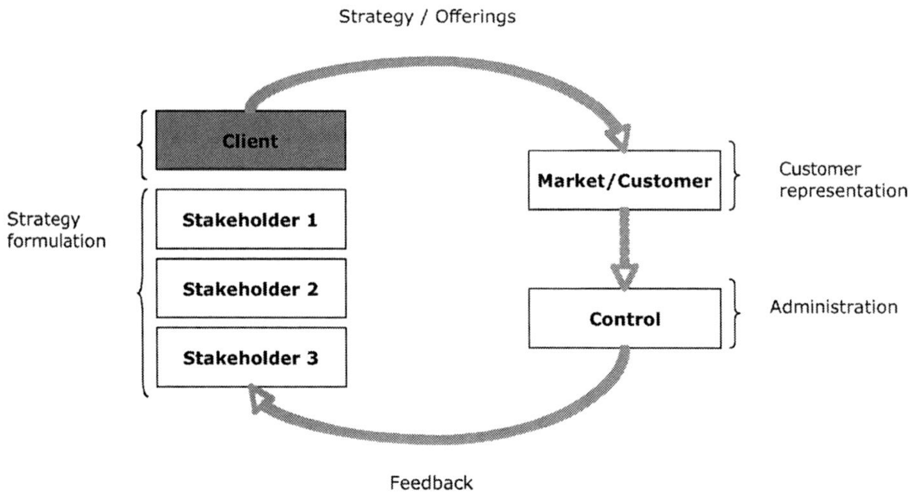

Figure 2. Move of a business wargame.

a single wargame: a client team, competitive or stakeholder teams, a market team and a control team. The first move occurs in the present and is based on data that is available to all of the teams. A move, as shown in the Figure 2, is a decision cycle, which begins with the stakeholders or competing companies and the client team taking first actions: competitive offerings, alliances, investments, or lobbying efforts.

The competitive or stakeholder teams will have to think about strategy, product, pricing, capacity, technology and other issues while considering the business environment. The client teams must execute and adjust their own strategy. An essential part of each move is the reaction of the customers, usually played by the market team, which provides the reaction of the customers, the size of the market, market segments, market share, revenue and how these figures change in the course of the moves. All these data will be passed on to the control team, usually run by the wargaming experts, who for instance use a financial model to calculate profits and losses. The control team is also in charge of supervising the wargame and introducing discontinuities (e.g. technological developments or policy issues) to add a touch of real-life dynamics (Fuld 2006). The control team can also assume the role of other stakeholders, such as regulators or competitors who are not represented in the game. This allows the control team to bring in new stakeholders during simulation, thus adding a certain dynamic to the stakeholder in a business wargame.

At the end of each move, the control team calculates the results of the competing or stakeholder teams' and the client teams' moves and uses these results as the starting position for the next move. While the market team reacts and the control team re-calculates the figures, the competitive teams and the client teams plan their next move before they learn the results of the previous move.

When describing how a business wargame works, it is important to make a clear differentiation between these kind of simulations (business wargaming) and computer-based simulations, and to discuss the role of the model used, with special attention to the unique character of business wargaming.

In reference to business wargames, Chussil (2007) argues that a model always underlines such a simulation, ranging from 'mental models', such as human judgement, to 'quantitative models'. However, the degree to which the model determines the simulation is relevant. The danger that models pose in guiding a simulation is that they represent the perspective of the analysts who constructed the model (Hanley 1992). This aspect is reflected, for instance, in a recent debate in the US Navy (Strategy Page 2008). While the US Navy relies on computer-based simulation for training purposes, it has been suggested that simulations, which are not as scripted as computer-based simulations, encourage the players to expect the unexpected. This form of simulation is also referred to as 'free wargaming' (Hanley 1992). Here the emphasis is put on the interaction among those individuals involved in the simulation without being excessively restricted by a computer-based model. It is essential that a business wargame be dynamic, driven by its participants and not, for instance, by a computer-based model with a set number of parameters, interlinked with pre-set sensitivities (Oriesek and Schwarz 2008). However, it should also be noticed that the underlying model, as pointed out ranging from mental models to simple financial models, always have the tendency to limit such a simulation, pointing out to the critical role models play here.

4.4. *Does business wargaming work?*

Before the relevance business wargaming might have for developing foresight is discussed, evidence will be presented in this section on the usefulness of this kind of simulation.

Taking a historic and military perspective, Hanley (1992) stresses that the military application of wargames helped participants to broaden the range of alternatives and to review their objectives. Minis and Tsamboulas (2008) present a case in which wargaming successfully tested contingency plans for the transport operations of the Athens 2004 Olympic Games. Scherpereel (2003, 70) reports that these simulations 'result in participants doing things differently, thinking longer term, seeing the big picture and better understanding the complexities of the competitive landscape'. His experimental study, quantifying the impact of a business wargame, supports this claim. Green (2002) presents another interesting piece of empirical evidence on role-playing – which can also be perceived as business wargaming in contrast with game theory – and discusses unassisted judgement in forecasting decisions in conflict situations.

However, are these arguments sufficient to support the claim that business wargaming can enable organisations to develop foresight, or that it even has advantages over other approaches? In an attempt to answer these questions it might be helpful to discuss what can be assessed as the goals of developing foresight and discussing how business wargaming can relate to them.

5. Developing foresight in a business wargame

This section will start out by assessing what can be perceived as the goals of developing foresight in an organisation, with the aim of discussing in more detail how business wargaming could contribute to achieving these goals. Further on, the development of foresight in a business wargame will be described in more detail.

5.1. *Learning and communication: how business wargaming relates to developing foresight*

In general, simulations in various formats can generate alternative pictures of the future (e.g. Bell 2003; Fuller and Loogma 2008). Further, this article has already argued that the purpose of a foresight process in an organisation is not so much to predict the future as to prepare for it (Tsoukas and Shepherd 2004). Foresight activities should be understood as structured communication processes focusing on mental models, blind spots and knowledge gaps. These processes should trigger discussions about present and future events that might have an impact, enabling an organisation to make sense of its environment and look for weak signals of change. In cognitive terms, a foresight process should enhance an organisation's 'memory of the future'.

Van der Heijden (1997) has put an emphasis on understanding scenarios, this emphasis is also relevant for foresight processes, as a tool for fostering and initiating strategic conversations in organisations. Following van der Heijden (1997), strategic conversation can be defined as a dialectical process of a scenario planning exercise that aims at forming a shared mental model of the organisation, its goals and the way in which it attempts to achieve them (Ratcliffe 2002). This definition implies that at the heart of foresight lays conjecture, and that at the heart of conjecture is conversation (Ratcliffe 2002).

Grant (2003, 511) states that Royal Dutch/Shell's scenario planning process was a vehicle for fostering organisational learning by sharing and integrating multiple knowledge bases from both within and outside the Shell group, with emphasis on communication and knowledge sharing: 'Shell's "scenario-to-strategy" framework involved discussion workshops in which scenarios would provide the foundation for an interactive strategy formulation'. Understanding a foresight process as a vehicle for strategic conversation in an organisation helps to define the purpose of such a process in that organisation. The value of such a strategic conversation is that mental models in an organisation are not only shared but also most likely to be challenged and that an organisation is likely to eliminate blind spots from its sensemaking of the environment and of the future. By referring to the participative and foremost interactive nature of a business wargame, it appears that what is happening in a business wargame relates to the claims of a strategic conversation.

Understanding a foresight process as a vehicle for strategic conversation is one possible role. The other role is that of learning, which can be related even more strongly to business wargaming. Perceiving a foresight process as a learning process (Costanzo 2004), allows for the argument that a foresight process can contribute to the 'learning organisation' (Senge 1990). This notion shall be underlined with reference to the concept of the 'memory of the future'.

When the future cannot be predicted and trends derived from a foresight process are likely to have at least one counter trend, why even bother with such an activity? The relevance of the concept of the 'memory of the future' (Ingvar 1985; De Geus 1997; Van der Heijden et al. 2002), has only recently been explored (Schacter, Addis, and Buckner 2007; Gilbert and Wilson 2007). Ingvar (1985) concludes that the human brain is able to store various pictures of the future. The more memories of the future that are stored, the more receptive an individual can be to signals from the outside world.

Wenzler and Chartier (1999, 380) argue that games and simulations, in the context of learning, allow 'organizations to test different approaches and solutions within a safe setting, thus helping them learn how to perform in the future. By facilitating such rehearsals of the future and by helping organizations envision and explore a multitude of time paths, games and simulations can effectively build their memories of future(s).' In the context of neuroscience, Schacter, Addis, and Buckner (2007) assert that event simulation has a key role in helping individuals plan for the future. In this article, one could argue that this is about developing foresight in a business wargame.

Van der Heijden et al. (2002, 177) reiterate the importance of the 'memory of the future' in reference to the scenario process, which is also applicable to a foresight process: 'In essence, the scenario process enabling managers to visit and experience the future ahead of time, thereby creating what is called 'memories' of the future. These visits to anticipated futures are then remembered, creating a matrix in the mind of managers and serving as subconscious guides to make sense of incoming environmental signals and to act on them'.

While computer-based simulation games are widely used for management education (Wolfe and Roberts 1993; Wellington, Faria, and Nulsen 1996; Mitchell 2004) there is an ongoing debate over whether or not the application of business simulations is of value to teaching (e.g. Doyle and Brown 2000). Doyle and Brown (2000) point out that critics question the validity of the simulation for teaching. However, the advantages of simulations in regard to other teaching methods can be summarised as follows: they are more realistic, more enjoyable (a clear prerequisite of learning), they encourage the development of team working skills, they take place in a risk-free environment, they allow participants to try out new behaviours and they encourage experimental and action learning (Doyle and Brown 2000). Wolfe and Rogé (1997) cite a study which concludes that business games are the most effective method of teaching strategic management, compared to cases, lectures and textbooks.

While it appears that business games or simulation can mainly be perceived as computer-based simulations, one can argue that by not limiting such games and simulations to a preset computer model, it is possible to expand the usefulness of these approaches. In this realm, Fuller and Loogma (2008) point out the advantages of role-playing simulations, which refer to business wargaming, over computer-based simulations. They argue that computer-based simulations seem incapable of producing social knowledge *per se*, whereas in a role playing simulation new meanings appear to be generated from the known through social interaction.

5.2. *Characteristics of business wargaming*

A typical business wargame is designed to evolve several years into the future. The participants are forced to think ahead, to evaluate their predictions and to question their mental models. In this process the participants experience several possible futures, not just one. In the course of a business wargame, the future dynamics of the stakeholders in a market or industry are analysed and anticipated. By simulating the future, participants 'experience' future dynamics and, consequently, develop foresight.

In addition to the future-orientation, a business wargame has several other characteristics, each of which appear to be responding to the criticisms levelled against scenario planning and to the obstacle of cognitive barriers.

A business wargame actively involves its participants. In the course of a business wargame, the dynamics of a market or competitors will be analysed and anticipated. By simulating an industry, market, or competition over time, managers are able to 'experience' the dynamics and the competition in their market or industry and to see the consequences of their strategy. Bracken (2001, 18) argues: 'The problem with many strategy techniques is that they are too cold and bloodless. They fail to capture human emotions, and because of their icy rational character, people don't really pay attention to them. They are soon forgotten, and they make no lasting impact on the organization. Gaming is profound learning experience, one that is not soon forgotten.'

Business wargaming, then, has the power to involve its participants not only during the simulation but also afterward. At this point the claim can be made, that business wargaming has the potential to bypass cognitive barriers. However, this aspect certainly would need to be further researched. Also, the argument can be made that a business wargame is a powerful tool to challenge the mental models of its participants: it forces them to step out of their usual roles in their organisations when they play competitors or other stakeholders; it uses the dynamic and complex nature of a business wargame to challenge one's perceptions of the future; and it uses a different setting to expose participants to ideas which might be counter to their mental models, thereby making it more likely that they will accept these new ideas.

As Bracken (2001) points out, a business wargame is a sustainable learning experience. Pina e Cunha, Palma, and Guimaraes da Costa (2006, 943) contend that managers are confronted with the tension between the need to know how the future might evolve and the fear of knowing that it might be contradictory to their mental models. While they go on arguing, that 'the foresight process offers an adequate ground to analyse the dialectical "back and forth relationship" which involves the need to know and the fear of knowing', these characteristics of a business wargame seem not only to support this notion, but also to amplify it.

Kurtz (2003) summarises the benefits of a business wargame as forcing a rigorous examination of the situation from several perspectives by actively involving the participants by providing them with multiple opportunities to learn. In addition, a business wargame can convey a deeper understanding of the competitive situation, how a participant's or a team's strategy will play out, and how the industry will develop in the future (Kurtz 2003). Dunnigan (2000) describes a wargame as an attempt to get a jump on the future and Fuld (2006, 109) adds: 'War games are not crystal balls but they do throw off lots of insight – and sometimes even foresight'.

In fact, a business wargame allows a management team not only to develop foresight, but also the *ex-ante* testing of hypothetical strategies. Testing a strategy in a business wargame will contribute to developing foresight, since a business wargame is usually designed to evolve several years into the future. Consequently, one could argue that when a strategy is supposed to be tested in a business wargame, foresight will be developed, while when the focus of a business wargame is to develop foresight, this can probably be very well achieved, when the simulation is centred on a strategy and the question how robust this strategy might be in the future.

In addition, following the proposition of Mendonça et al. (2004) that foresight is a socio-technological approach from which organisational improvisation can emerge, the relevance of foresight in the context of business wargaming becomes even more evident. Mendonça et al. (2004, 210) state: 'In practise, improvisation is said to exist when deliberate but unplanned actions are executed, aiming to help the organisation in the effort to respond to unforeseen opportunities or threats in an environment under perpetual change'. With the combination of developing foresight and testing a strategy, business wargaming provides a frame in which organisations can actually test and develop their improvisation capabilities concerning unforeseen threats or opportunities and actually learn from errors.

The characteristics of business wargaming can be summarised:

- It is future-oriented.
- It encourages the development of foresight.
- It is dynamic and involves active anticipation.
- It focuses on the market, the competitors and their dynamics, as well as on the company itself.
- It teaches participants to identify early signals of change.
- It enables participants to discover weaknesses in their strategies.
- It allows testing strategies prior to implementation.
- It allows testing and developing organisational improvisation skills.

5.3. *Integrating business wargaming into a foresight process*

How can business wargaming add to a foresight process (Figure 3)? While a business wargame cannot replace a constant foresight process in an organisation, it can add to and improve the company's foresight ability. A business wargame can identify weak signals that can be used as a starting point for a foresight process.

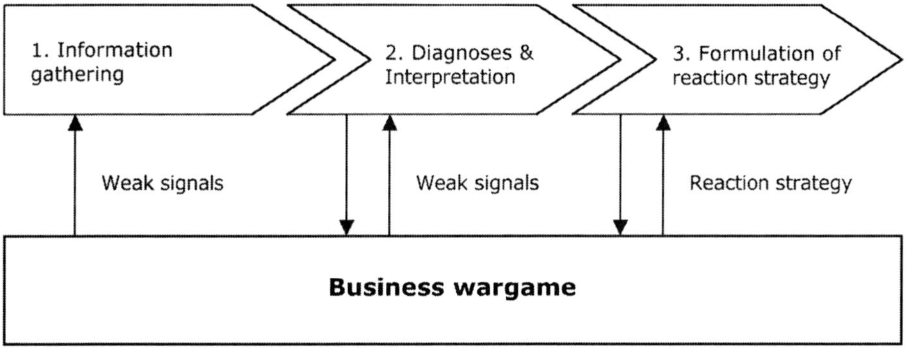

Figure 3. Integrating business wargaming into the foresight process.

The diagnoses and interpretation phase of the foresight process can be benefited if weak signals or relevant trends have been identified in a business wargame. Conversely, if weak signals have been detected and if the task is to diagnose and interpret them, these weak signals can be injected into a business wargame so they can be diagnosed and interpreted in the context of a dynamic simulation. The third phase of the foresight process, the formulation and testing of reaction strategies, can also be achieved within a business wargame, testing, for instance a potential reaction strategy. The insight gained during a business wargame, of course, can add to the third phase of a foresight process.

Business wargames can be of value when they complement a foresight process. However, even if an organisation has not set up a process to develop foresight, a business wargame can be of value in developing foresight.

5.4. *Critical aspects of developing foresight in a business wargame*

After having discussed the potential relevance business wargaming can have for developing foresight, it appears to be inevitable to discuss some critical aspects of this approach.

As a more recent discussion on scenario planning (Hodgkinson and Wright 2002, 2006; Whittington 2006a, b) has demonstrated, how these kind of exercises are carried out, how they are prepared and designed, play a major role if the goals of an organisation are to develop foresight. Further, if a business wargame is more or less cost extensive as, for instance, a scenario exercise or other foresight activities depends certainly on the kind of business wargame and its purpose. However, the preparatory work for business wargaming is quite time consuming, since the profiles of each represented stakeholder and the industry itself in which the wargame is taking place need to be researched very carefully. Of course, this can be a very valuable process, especially if an organisation lacks a sufficient competitive intelligence (Gilad 2004) function. Widmer (1980) refers to a quote from Frederick the Great, emphasising the relevance of competitive intelligence and that strategic error tends to result from false premises: 'To plan a successful campaign you must know your enemy, your allies, and the nature of your terrain'. In addition, the participating managers need to invest their time, which due to the nature of such a simulation is of paramount importance. Overall, a business wargame demands a lot of resources from an organisation.

When is it appropriate to apply business wargaming for developing foresight, when should an organisation invest in such an activity? Clearly, and this point has been emphasised here already, when the competitive environment of an organisation is involved and foresight is needed on aspects such as how the competitive environment will change or how new potential competitors will change the industry, then the application of a business wargame appears to be useful. In general, business wargaming appears to be helpful when the perspectives of other stakeholders need to be involved in developing foresight. In addition, one could argue that because a business wargame usually evolves into the future, developing foresight takes place in nearly every business wargame and is one of the unique features of this kind of simulation.

6. Conclusion

Wargaming is well established in the military field and on the rise in management. It appears that business wargaming can be of value for developing foresight because of its dynamic and participative character. It helps participants to overcome the cognitive barriers, which tend to hinder an organisation's foresight processes and activities and helps companies to anticipate possible actions of competitors by experiencing the perspective of those competitors. However, more research on business wargaming is needed. At this point one could argue, that the theoretical as well as the empirical basis are not very strongly developed. In particular, it seems to be of relevance to research how business wargaming can help managers to challenge mental models and overcome cognitive barriers, but also the question should be raised how and eventually if business wargaming can contribute in an organisation to the acquisition of knowledge?

The framework described in this article suggests that a business wargame can add to a foresight process and vice versa. While a business wargame should complement a foresight process, a wargame can also be a vehicle through which to develop foresight in a single exercise, or even to mark the starting point for a continuous foresight process. By involving its participants actively in a dynamic strategic simulation, business wargaming has the potential to challenge mental models, foster learning and develop the kind of foresight that is essential to success in an increasingly dynamic and complex business environment.

References

Aguilar, F.J. 1967. *Scanning the business environment*. New York: Macmillan.

Andlinger, G.R. 1958. Looking around: what can business games do? *Harvard Business Review* 36, no. 4: 147–152.

Ansoff, H.I. 1975. Managing strategic surprise by response to weak signal. *California Management Review* 18, no. 2: 21–33.

Bazerman, M.H., and M.D. Watkins. 2004. *Predictable surprises*. Boston, MA: Harvard Business School Press.

Bell, W. 2003. *Foundations of futures studies*. New Brunswick: Transaction Publishers.

Bellman, R., C.E. Clark, D.G. Malcolm, C.J. Craft, and F.M. Ricciardi. 1957. On the construction of a multi-stage multi-person business game. *Operations Research* 5, no. 4: 469–503.

Bracken, P. 2001. Business war gaming. *Scenario & Strategy Planning* 3, no. 2: 15–18.

Chia, R. 2004. Re-educating attention: what is foresight and how is it cultivated? In *Managing the future*, ed. H. Tsoukas and J. Shepherd, 21–37. Oxford: Blackwell Publishing.

Chussil, M. 2007. Learning faster than the competition: war games give the advantage. *Journal of Business Strategy* 28, no. 1: 37–44.

Costanzo, L.A. 2004. Strategic foresight in a high-speed environment. *Futures* 36, no. 2: 219–235.

Courtney, H. 2001. *20/20 Foresight: crafting strategy in an uncertain world.* Boston, MA: Harvard Business School Publishing.

Day, G.S., and P.J.H. Schoemaker. 2004. Driving through the fog: managing at the edge. *Long Range Planning* 37, no. 2: 127–142.

Day, G.S., and P.J.H. Schoemaker. 2006. *Peripheral vision: detecting the weak signals that will make or break your company.* Boston, MA: Harvard Business School Press.

De Geus, A. 1997. *The living company.* Boston, MA: Harvard Business School Press.

Doyle, D., and F.W. Brown. 2000. Using a business simulation to teach applied skills – the benefits and the challenges of using student teams from multiple countries. *Journal of European Industrial Training* 24, no. 6: 330–336.

Dunnigan, J.F. 2000. *Wargames handbook.* New York: Writers Club Press.

Fahey, L., and R.M. Randall. 1998. *Learning from the future: competitive foresight scenarios.* New York: John Wiley.

Fuld, L. 2006. *The secret language of competitive intelligence.* New York: Crown Business.

Fuller, T., and K. Loogma. 2008. Constructing futures: a social constructionist perspective on foresight methodology. *Futures*, doi:10.1016/j.futures.2008.07.039.

Gharajedaghi, J. 1999. *Systems thinking.* Boston, MA: Butterworth Heinemann.

Gilad, B. 2004. *Early warning: using competitive intelligence to anticipate market shifts, control risk, and create powerful strategies.* New York: AMACOM.

Gilbert, D.T., and T.D. Wilson. 2007. Prospection: experiencing the future. *Science* 317, no. 7: 1351–1354.

Ginter, P.M., and A.C. Rucks. 1984. Can business learn from wargames? *Long Range Planning* 17, no. 3: 123–128.

Grant, R.M. 2003. Strategic planning in a turbulent environment: evidence from the oil majors. *Strategic Management Journal* 24, no. 2: 491–517.

Green, K.C. 2002. Forecasting decisions in conflict situations: a comparison of game theory, role-playing, and unaided judgement. *International Journal of Forecasting* 18, no. 3: 321–344.

Haffa, R.P., and J.H. Patton. 1999. The need for joint wargaming: combining theory and practice. *Parameters*, Autumn: 106–118.

Hamel, G., and C.K. Prahald. 1994. *Competing for the future.* Boston, MA: Harvard Business School Press.

Hanley, J.T., Jr. 1992. On wargaming: a critique of strategic operational gaming. PhD diss., Yale University.

Helmer, O. 1983. *Looking forward: a guide to futures research.* London: Sage.

Hodgkinson, G.P., and G. Wright. 2002. Confronting strategic inertia in a top management team: learning from failure. *Organization Studies* 23, no. 6: 949–977.

Hodgkinson, G.P., and G. Wright. 2006. Neither completing the practice turn, nor enriching the process tradition: secondary misinterpretation of a case analysis reconsidered. *Organization Studies* 27, no. 12: 1895–1901.

Horton, A. 1999. Fore front: a simple guide to successful foresight. *Foresight* 1, no. 1: 5–9.

Ilmola, L., and O. Kuusi. 2006. Filters of weak signals hinder foresight: monitoring weak signals efficiently in corporate decision-making. *Futures* 38, no. 8: 908–924.

Ingvar, D.H. 1985. Memory of the future: an essay on the temporal organization of conscious awareness. *Human Neurobiology* 4, no. 3: 127–136.

Kalman, J.C., and E. Rhenman. 1975. The role of management games in education and research. In *Gaming-simulation: rationale, design, and application*, ed. C.S. Greenblat and R.D. Duke, 233–269. New York: Sage.

Krystek, U., and G. Müller-Stewens. 1993. *Frühaufklärung für Unternehmen: identifikation und Handhabung zukünftiger Chancen und Bedrohungen [Early warning for corporations: identification and management of future opportunities and risks].* Stuttgart: Schäfer-Poeschel.

Kurtz, J. 2003. Business wargaming: simulations guide crucial strategy decisions. *Strategy & Leadership* 31, no. 6: 12–21.

Liebl, F. 1996. *Strategische Frühaufklärung: Trends – Issues – Stakeholders [Strategic early warning: trends – issues – stakeholders].* München: R. Oldenburg Verlag.

Liebl, F. 2000. *Der Schock des Neuen: entstehung und Management von Issues und Trends [Shock of the new: emergence and management of issues and trends].* München: Gerling Akademie Verlag.

Liebl, F. 2004. Organizational foresight as strategic knowledge management: a frame of reference. Paper presented at the 2nd International Conference on 'Organizational Foresight' at the University of Strathclyde, Graduate School of Business, Glasgow, Scotland, UK.

MacKay, R.B., and P. McKiernan. 2004. Exploring strategy context with foresight. *European Management Journal* 1, no. 1: 69–77.
Makridakis, S. 2004. Foreword: foresight matters. In *Managing the future*, ed. H. Tsoukas and J. Shepherd, J., XIII–XIV. Oxford: Blackwell Publishing.
Martin, B.R., and R. Johnson. 1999. Technology foresight for wiring up the national innovation system: experiences in Britain, Australia, and New Zealand. *Technological Forecasting and Social Change* 60, no. 1: 37–54.
Mendonça, S., M. Pina e Cunha, J. Kaio-Oja, and F. Ruff. 2004. Wild cards, weak signals and organizational improvisation. *Futures* 36, no. 2: 201–218.
Miles, I., M. Keenan, and J. Kaivo-Oja. 2003. *Handbook of knowledge society foresight*. Dublin: European Foundation for the Improvement of Living and Working Conditions.
Minis, I., and D.A. Tsamboulas. 2008. Contingency planning and war gaming for the transport operations of the Athens 2004 Olympic Games. *Transport Reviews* 28, no. 2: 259–280.
Mitchell, R.C. 2004. Combining cases and computer simulations in strategic management courses. *Journal of Education for Business* 79, no. 4: 198–204.
Müller, G. 1981. *Strategische Frühaufklärung* [*Strategic early warning*]. München: Planungs- und Organisationswissenschaftliche Schriften.
Neugarten, M. 2006. Foresight – are we looking in the right direction? *Futures* 38, no. 8: 894–907.
Oriesek, D.F., and R. Friedrich. 2003. Blick in die Zukunft [View into the future]. *Harvard Business Manager*, May: 65–71.
Oriesek, D.F., and J.O. Schwarz. 2008. *Business wargaming: securing corporate value*. Aldershot: Gower Publishing.
Perla, P.P. 1990. *The art of wargaming*. Annapolis, MD: Naval Institute Press.
Pina e Cunha, M., P. Palma, and N. Guimaraes da Costa. 2006. Fear of foresight: knowledge and ignorance in organizational foresight. *Futures* 38, no. 8: 942–955.
Pohl, F. 1996. Thinking about the future. *The Futurist* 30, no. 5: 8–13.
Postma, T.J.B.M., and F. Liebl. 2005. How to improve scenario analysis as a strategic management tool. *Technological Forecasting & Social Change* 72, no. 1: 161–173.
Ratcliffe, J. 2002. Scenario planning: strategic interviews and conversations. *Foresight* 4, no. 1: 19–30.
Reading, A. 2004. *Hope & despair*. Baltimore: The Johns Hopkins University Press.
Schacter, D.L., D.R. Addis, and R.L. Buckner. 2007. Remembering the past to imagine the future: the prospective brain. *Nature Reviews Neuroscience* 8, no. 9: 657–661.
Scherpereel, C.M. 2003. The impact of business war games: quantifying training effectiveness. *Development in Business Simulation and Experiential Learning* 30: 69–82.
Schoemaker, P.J.H. 1992. How to link strategic visions to core capabilities. *Sloan Management Review* 34, no. 1: 67–81.
Schwarz, J.O. 2005. Pitfalls in implementing a strategic early warning system. *Foresight* 7, no. 4: 22–30.
Schwarz, J.O. 2006. *The future of futures studies: a Delphi study with a German perspective*. Aachen: Shaker.
Schwarz, J.O. 2008. Assessing the future of futures studies in management. *Futures* 40, no. 3: 237–246.
Seidl, D., and D. van Aaken. 2004. Anticipating critique and occasional reason: modes of reasoning in face of radically open future. Paper presented at the 2nd International Conference on 'Organizational Foresight' at the University of Strathclyde, Graduate School of Business, Glasgow, Scotland, UK.
Senge, P.M. 1990. *The fifth discipline: the art & practice of the learning organization*. London: Century Business.
Strategy Page. 2008. US Navy runs from reality. http://www.strategypage.com/htmw/htlead/articles/20080605.aspx.
Treat, J.E., G.E. Thibault, and A. Asin. 1996. Dynamic competitive simulation: wargaming as a strategic tool. *Strategy, Management, Competition*, Second Quarter: 46–54.
Tsoukas, H. 2004. Coping with the future: developing organizational foresightfulness. *Futures* 36, no. 2: 137–144.
Tsoukas, H., and J. Shepherd. 2004. Introduction: organization and the future, from forecasting to foresight. In *Managing the future*, ed. H. Tsoukas and J. Shepherd, 1–19. Oxford: Blackwell Publishing.
Van der Heijden, K. 1997. *Scenarios: the art of strategic conversation*. Chichester: John Wiley.
Van der Heijden, K., R. Bradfield, G. Burt, G. Crains, and G. Wright. 2002. *The sixth sense*. Chichester : John Wiley.
Wellington, W., A.J. Faria, and R.O. Nulsen, Jr. 1996. An empirical investigation into the nature of the learning process in a computer-based simulation game. *Marketing Education Review* 6, no. 3: 15–28.
Wenzler, I., and D. Chartier. 1999. Why do we bother with games and simulations: an organizational learning perspective. *Simulation & Gaming* 30, no. 3: 375–384.
Whittington, R. 2006a. Completing the practice turn in strategy research. *Organization Studies* 27, no. 5: 613–634.
Whittington, R. 2006b. Learning more from failure: practice and process. *Organizational Studies* 27, no. 12: 1903–1906.

Widmer, H. 1980. Business lessons from military strategy. *McKinsey Quarterly* 2: 65.
Wolfe, J., and C.R. Roberts. 1993. A further study of the external validity of business games: five-year peer group indicators. *Simulation & Gaming* 24, no. 1: 21–33.
Wolfe, J., and J.N. Rogé. 1997. Computerized general management games as strategic management learning environments. *Simulation & Gaming* 28, no. 4: 423–441.

Weak signals as a flexible framing space for enhanced management and decision-making

Pierre Rossel

College of Management of Technology, Swiss Federal Institute of Technology in Lausanne (EPFL), Switzerland

This paper revisits theories of weak signal analysis using the concepts of 'framing' and 'meta-framing' to provide an enriched methodological environment for management and decision-making in turbulent contexts. This attempt takes place within the effort that futures studies have deployed to cope with the need to be 'early' in change processes, and is intended to improve firms' preparedness regarding uncertainties, threats and opportunities. The weak signal notion is one of the metaphors coined to accomplish that, thanks to the seminal role played by Igor Ansoff in the 1970s and 1980s, and the inputs of more recent continuators who have shaped a wide methodological environment. After over 20 years of weak signal exploration, some pending issues suggest the need to re-open some basic hypotheses as well as the use of a wide-angle lens to review the Ansoff legacy and its relevance for management and decision-making. The concepts of framing and meta-framing are the means that will allow us to do this by adding a more reflexive dimension to weak signal analysis, taking account of analysts' and commissioning organisations' own biases, and enhancing firms' robustness against possibly disruptive futures. Pragmatic managerial orientations of how to make use of methodological blends involving reflexive steps will be outlined, also suggesting diverse tracks for further research on weak signal theories and their contributions to both futures studies and strategic management.

1. 'Weak signal', a metaphor to tackle the need to be 'early' regarding changes and uncertainties

The history of futures studies is already quite substantial. Each generation of futurists had to cope with particular historical and epistemological challenges, appraised as being more or less successful and relevant by futurists of the following generations. For quite some time, long-term foresight issues (i.e. anything from 20 to 300 years ahead) have been fashionable. Shorter-term concerns were somehow embedded in longer-term development as parts of forecasting or system dynamic outcomes. It took the visionary efforts of some analysts to raise shorter-term futures, or shorter anticipation and response time capabilities regarding future changes, as theoretical challenges in their own terms.

In that evolution of the discipline, let us emphasise the prominent role of Igor Ansoff, pioneering the field in the 1970s and 1980s[1] and some 15–20 years later (between 1995 and the time of writing – 2008) that of a larger group of analysts, with a clear stronghold in the Finnish arena of futures studies. These more recent contributors explored further the weak signal issue, but essentially building upon Ansoff's work, always displaying a high respect for weak signal theory's roots.

Without putting too much emphasis on the critique tone, this paper aims at examining options that still remain open for discussion, as a way to re-enforce the weak signal approach. We will thus go across the contributions of Ansoff and others, and pinpoint strengths and original features, but also, in a further section, some weaknesses. We will propose conceptual solutions to make the mainstream approach to weak signal management more robust, envisaging a diversity of options and methodological blends hopefully enriching firms' 'futures' sensitivity.

The need to be 'early' on processes of change has been handled so far mainly through the concept of weak signal and to a lesser extent by the notion of 'early warning' (see for that, Gilad 2003), as well as neighbour ideas which have flourished in other languages, such as 'Früerkennung' or 'Frühauflärung' in German (early detection, early insight) or 'les faits porteurs d'avenir' (facts they may help shape a new future) in French. Let us not forget 'wild cards', which in futures studies usually stand as tokens for surprises, and which are often treated as near equivalent of, or as having tight relationships with weak signals (weak signals as advanced forms of wild cards or wild cards as being simply a particular kind of weak signals).[2] In this conceptual paper, the broad attempt at being 'early' in detecting changes will be analysed through the notion of weak signal. More suggestions will be made in the concluding section.

As we shall see in Section 4, the first thing to remember is that the notion of weak signal does not necessarily correspond to a real event or process, but is only a metaphor, building upon information theory to help us tackle a hypothetical relationship that we, as individuals, organisations or societies, develop with our environment and its emerging characteristics. Most of this relationship passes by as a series of perceptions, representations and models of how the world functions, not only physically but also in socio-economic, cultural and political terms. This activity is, most of the time, routinely carried out and tends to stabilise, for a given period of time, on certain paradigms (or dominant models) that somehow shape this perceptive and representational capability at both the micro- and macro-levels of our reality. What we call a 'weak signal' is usually not a direct warning message but, at best, a series of perceptions of issues, events or processes that might constitute an indication of further developments, possibly impacting our lives in the future.

In order to understand more fully the originality of such a relationship between the present and the future, let us examine what some established definitions of weak signals reveal. Ansoff formulated the idea that they are symptoms of possible change in the future. This broad perspective was presented some years later by Coffman (1997), who paid a more detailed attention to conditions and stages of evolution of weak signals. According to him, a weak signal is an idea or trend with potential impact on a business or its environment, which may be new or surprising from the signal receiver's vantage point (a perception which may differ for others), thus making it difficult to identify and isolate from other less relevant indications; this difficulty explains in particular why some people may scoff at weak signal suggestions, which are in fact indicators for firms and organisations. It is worth mentioning the proposal of Kamppinen, Kuusi, and Söderland (2002), who see in a weak signal an individual event or a group of inter-related events, which may not seem important initially, but develop into significant phenomena in shaping the future. Finally, let us also stress the interesting comment made by Mannermaa (2004) reported by Hiltunen (2007), who relates how the seemingly small aspects of reality represented by weak signals can be connected to big impacts later on.

Drawing upon these proposals, and bearing in mind firms' management and decision-making needs, we see three key issues worth examining:

(1) 'Surprise for some, not for others': it is important to understand how this can be. Is the reason for such a variation lying in the intelligence of some 'above average' players as opposed to less smart ones? Or is there another problem, still to be identified, which can justify this statement? We will attempt to answer this question in stages.
(2) 'Weak then strong', but also 'small then big': this makes sense, as faintly perceptible events or facts that can grow and then have a big impact are the only ones which should retain a firm's strategic attention. As a matter of fact, small signals with small impacts are of no interest as it happens all the time; big signals with small impacts can be, in principle, neglected; big signals with big impacts correspond, most likely, to processes which should have been cared for earlier (fitting the strong signal profile). The only other option worth considering is the gradual increase of the signalling strength, along with gradual increase of impact, but this is typically the situation that strategic planning is supposed to help handle, corresponding to a monitoring or dashboarding activity that any enterprise in its market struggle must continue. To sum up, the 'small then big' sequence is the only weak signal profile that management should focus on.
(3) Patterns of evolving 'from weak to strong' signal: If there are numerous models of change, the question is which ones fit best the analysis of weak signal paths and, more generally speaking, the early detection scheme? Among the main patterns likely to satisfy a broad usage of the weak signal notion, it is convenient to make a distinction between expected forms of the change process, where weak signals are only the first occurence of a typical signature (first snowfall for the yearly coming of winter for instance) and completely open-ended dynamics, for which we have no clear idea of how they will develop and for which, of course, weak signals are also more difficult to envisage. Another useful distinction to stress is the opposition between narrow, strict and well-documented causation models and more hybrid or complex aggregations of change processes. As an example of a narrow causation, let us mention the deterioration of traditional food habits in regions until now little touched by the fast food trend and its multiple health consequences; as an illustration of an open-ended and complex process, let us consider the case of mobile telecommunications development, which is much more than just transportable telephones. Finally, the speed of change can also vary quite a lot, with gradual and even-paced processes on the one side and various forms of accelerations on the other, mainly exponential (the Google type) or spousing the sudden pace shift. In this particular category, there are several possible sub-patterns, including the famous 'butterfly effect', which is quite important for the modelling of climate change, as well as in the financial domain.

Beyond these generic issues, however, there are still some fundamental questions to address: are weak signals real things or particular perceptions? How to sort out the good ones from the less significant ones? Does the fact that weak signals can be surprises only to some people come rather from insufficient information, bad modelling or unsuitable methodologies? Did the enhancements proposed by Ansoff's more recent continuators overcome most of the problems to be tackled (strategic surprises), or should we consider that some important issues are still pending? These are the questions we want to address and for which tentative answers will be provided in the following sections.

2. A strong theoretical programme pursued over three generations of futurists

2.1. *Ansoff's seminal contribution*

Igor Ansoff was already a star in the field of management when he proposed new concepts for tackling uncertainties and surprises. He was for instance the first one to coin the concept of strategic planning. During the 1970s and the 1980s, he developed a complementary set of theories and methodological tools to cope with what he called 'turbulent environments', 'strategic discontinuities' and 'strategic surprises'. Before that, futures studies were dominated by ideas that we could somehow deduce the future from past and present. Main approaches were then: (1) forecasting, envisaged in the strict sense of calculating possible trajectories based upon past series of data; (2) Delphi exercises, making good use of multi-expert knowledge to put next steps of a given problem into perspective; (3) simulations, based upon some form of systems thinking; and (4) scenario developments, often combining several of these methods. With Ansoff's new conceptual proposal, the idea emerges that the past is not always a fast track to envisage futures, at least not as the result of incremental and continuous development. As soon as we leave the past behind, many things can happen which may surprise even the most renowned specialists. In reality, Ansoff does not get totally away from the idea that the past is important to understand the future. He only suggests that some futures have, at present, little evidence of what they may become or help shape, such that they are often hard to identify. Hence, the notion of weak signal that an organisation must learn how to capture and make good use of, before the signal becomes too strong.

Let us thus examine the essentials of Ansoff's contribution (in particular 1975, 1982, and 1984), envisaged not chronologically but as a meaningful whole. Ansoff developed the convenient and yet original idea that, in our selective and interpretative activities, we tend to introduce filters, that help us make sense of what we perceive and take action accordingly. He thus imagined applying three types of filtering operations so as to transform the information flow of perceived weak signals into usable knowledge. In fact, what he had conceived for the strategic management of organisations was something that we all do normally but subconsciously; but of course, in Ansoff's perspective, this mundane habit is deployed within a consistent methodological framework. The almost unnoticed filters of our daily lives have thus become, with Ansoff, a rather ambitious programme for selecting and interpreting emerging signals and the related decision-making and strategic endeavours. Let us summarise this already well-known process as follows:

For an organisation wanting to cope with uncertainties and possible strategic surprises, any clue from its environment, such as a new phenomenon or event, in order to become a meaningful reality, has to pass through three different filters; at the end, the objective is to have effective decision-making options.

(1) The observation filter (or surveillance filter) defines the area that we can observe and collect data.
(2) The cognitive filter (also called mentality filter) is used for evaluating the novel issue that has passed the observation filter, and in particular its relevance for the firm.
(3) The power filter is engaged when it comes to involving decision-makers and their particular experience and interest, as well as any players who has been in the analytical process for a while and having a role to defend.

This filtering mechanism is quite expressive, but by itself insufficient. Ansoff provides us with a complementary framework that shows how the organisation can gradually cope with weak signals, suggesting a life cycle of how problems emerge and become noticeable. Also, in a second stage, what type of attitude and, finally, what response should be given to each successive level of manifestation.

Beside this progressive perspective on how weak signals can turn into strong ones and nevertheless be effectively managed by an organisation, Ansoff views the firm's responses as being tightly related to the state of knowledge. He thus divides the responses into two main groups: responses affecting the firm's relationship with the environment and responses modifying the internal dynamics and structure of the firm. For both of them, he sees three increasingly stronger strategies: one that improves the firm's awareness and understanding, one that increases the firm's flexibility and one that allows the firm to directly tackle threats or opportunities.

Although the response time of the firm is always given by its initial level of preparedness and flexibility, the idea of progressive shaping is key to Ansoff's handling of weak signals. It has been presented as the interplay between the firm and its changing environment, the outcome of which is translated into weak signals but also, increasingly, stronger signals. Formulated in knowledge terms, this gradual construction involves the successive stages of: (1) sense of a threat or opportunity, (2) increased knowledge of sources, necessary responses and possible outcomes, and (3) awareness from the firm that it must act, and not let signals become too strong; otherwise the firm may be too late in recognising new trends or discontinuous developments.

2.2. *Further contributions*

Although Ansoff developed throughout the years an impressive framework for taking weak signals into account in the strategic activity of firms, some academics and consultants have tried to improve his theory. Tracing back all their contributions would be a survey in itself, but some basic orientations can be identified, giving us a broad appraisal of how attractive Ansoff's legacy has been, and also how diversified the weak signal perspective has now become, a phenomenon that Ilmola and Kuusi (2006) see as a 'Renaissance' of weak signal analysis. Let us mention in particular the following inputs:

- A more precise description of the links between weak signals and information theory (e.g. Coffman 1997)
- The making of tools to facilitate weak signal generation and analysis (see Fountain Park[3])
- Multi-criteria analysis as a broader perspective for weak signal management (see Könnölä, Brummer, and Salo 2005)
- A fully-fledged information life-cycle view making sense of weak signals (see, e.g. Choo 2007), starting with such key stages as idea creation and elite awareness
- The sign as semiotic interplay: referring to the sign as a substitute for real life events and possible weak signals. This approach allows for distinguishing between endo- and exo-signs, and give a key role to the intervention of 'interpretants' for the decisive sense-making steps (see for that, Hiltunen 2008)
- Classification matrices: they aim at proposing a convenient and suggestive 2 × 2 dimensional mapping of weak signals. Let us mention in particular the attempt of Morrison and Wilson (1996), who distinguish between low/high certainty or uncertainty and low/high impact (a suggestion also close to Mannermaa 2004) and Kuosa's perspective (2005), who proposed confronting signals that are connected to a known trend or on the contrary to no known trend, to events which are to be differentiated between partly to completely expected

- Weak signals and wild cards: several kinds of relation are envisaged between these two concepts, from authors who envisage them as broad synonyms to others who consider that wild cards are weak signals that we cannot ordinarily think of (see Mendonça et al., 2004) and have to be generated thanks to specific procedures (Petersen 1999)
- Systemic transactions: in recent years, some authors have re-enforced the relationship of weak signal with system analysis, pursuing an already established tradition (Coffman 1997); most interesting in that domain is the work deployed by Snowden and colleagues (2002, 2005) to value particular configurations, allowing situations to evolve from specific transitional states of organisation to others; a creative perspective open to promising, as well as dangerous, outcomes to take care of

Other significant efforts continuing Ansoff's contribution are the various suggestions for finding weak signals. Mendonça et al. (2004) provide indications of where to look for, starting with the most obvious targets, while Steinmueller (2004, 2006) and Hiltunen (2007),[4] emphasise the importance of diverse social actors including those living at the fringes of society (social movements, artists, sexual minorities, 'deviants', youth, etc.). This goes in the same direction as Day and Schoemaker's (2006) recommendation to consider weak signals and change management issues by paying great attention to what they call the 'Periphery' (another interesting metaphor).

3. Pending problems

3.1. *A constructivist critique of the Ansoffian tradition*[5]

Examining the literature on weak signal, one is struck by how most authors end up forgetting that the concept they use is only a metaphor. As a matter of fact, in most cases, no one is sending signals to us. The same can be said of companion concepts such as filters and noise. The trend, so far, has been to 'naturalise' weak signals,[6] as if they were objects or features in their own right, waiting to be discovered, instead of considering them as the expression of the paradigmatic capacity of the analyst, whoever it may be, to organise perception and interpretation in a certain way. In the Ansoffian tradition, weak signals are portions of reality isolated as meaningful, picked up to become part of an interpretation, and then, eventually, strategic activity. The process through which these things called weak signals by analysts actually come into existence and are then turned into useful information for individuals or organisations is simply never envisaged as a research problem. In other words, habits, models, assumptions and models for constructing weak signals out of billions of other possible perceptions is not part of the analysis. What makes someone decide, without knowing the future, that a particular weak signal is likely to have a more or less certain fate, a greater or lesser impact or tighter or looser connections to known trends (see comment on the mapping of weak signal proposed by Morrison and Wilson 1996; Mannermaa 2004; and Kuosa 2005), is somehow black-boxed.

In a constructivist perspective, it is just the opposite: what makes us identify some features as weak signals and evolve with them, is indeed an important issue. In particular, there is a need to better understand how our goals, interests, expectations, fears or aspects of our training background, not to speak of the different styles of maintaining oneself in power and embedded in one's culture or psychology, may orientate our selection and bias our interpretation of weak signals. Yet, in spite of the so-called 'cognitive filter', there is hardly any provision in the weak signal literature for a reflexive thinking about the path to one's own findings or their further treatment. Expressed in constructivist terms, all the weak signals suggested by the Ansoffian perspective are merely 'candidate weak signals', hypotheses or starting points for an exploratory reflection, for firms wanting to cope with disruptive problems, changing environments and innovative forms of learning. This deficit of reflexive precautions in the analysis of early stages of change has several consequences.

3.2. *Vision troubles, knowledge management or complex issues*

The first problem is the vulnerability of an organisation which finds all sorts of weak signals, works on them, but is unable to question itself on the 'whys' and 'hows' of its choices. In the absence of such a reflexive stage, finding weak signals may just be a way of producing false positives or false negatives (just as working through Ansoff's gradual uptake of weak signals is no guarantee of success). One good way to understand this paradox is Holling's image of Moebius loops (2004), in which one may have the impression of coping with continuous change signals, but having, in reality, a crucial difficulty escaping a basic problem of entrenchment. The financial crisis, due to the collapse of the US subprime market since autumn 2007, through which all risk experts (bankers and quotation institutes in particular) have ended up converging towards the same beliefs and mutually re-enforcing choices, is a good example of that problem. The US$50 billion fraud by Bernard Madoff, ex-president of Nasdaq, also involved the indulging and friendly attitude of the US Securities and Exchange Commission. These examples have demonstrated how specialists can be inhibited by their own desires, as well as cumulative and converging (but false positive) signals and beliefs.

These issues have important managerial consequences. Some firms or industrial sectors may recognise a salient feature, but fail to take it into account as important. For instance, Polaroid was able to observe the arrival of digital photography but failed to perceive how disruptive it would be to the company. Similarly, the Swiss watch industry of the 1970s produced the first electronic watch but failed to envisage its market significance and lost two-thirds of its manpower in the 1980s. In general, disruptiveness (but that is what weak signals are for) is difficult to perceive because it is dissonant to our dominant mode of thinking, in the sense explored by Festinger (1957), a problem also diversely emphasised by Levinthal and March (1993), Mannermaa (2004) and Ilmola and Kuusi (2006).

This myopia may be increased by the overall importance of the power filter, to take up Ansoff's concept. The question was originally raised as to how firms' leaders could trigger an internal process of identifying and then interpreting, and finally enacting, weak and progressively stronger signals, but many studies show that the reasons why a firm would start a foresight exercise or maintain a high level of alertness regarding change can be quite diverse (see Gruber et al. 2003). Freeman and Pattinson (2007), for instance, have examined the various strategies behind the start of a foresight exercise and also the forms of interferences that top manager introduce in the foresight exercises which they commission. Far from being open-ended, it is not infrequently biased. This observation, based on consistent fieldwork and findings, should not be viewed as a mistake or lack of intelligence. It reflects, rather, how difficult it is to support a thoroughly open reflection on changes to come. In this sense, one has to recognise that Ansoff's various filtering steps are not equal, the power filter playing often the most important role, dictating back how the other two filters of the Ansoff script should be conceived and applied (and also how weak signals are selected).

This bias limits the importance of the 'knowing vs not knowing' issue (having access to the 'perfect information' or not). Whistle-blowers of the Challenger space shuttle catastrophe were fired (see for other examples, Morel 2002); in this case, the power filter was indeed the key problem (just as in the Chernobyl case, there was no room for risk awareness). Without some reflexive moves, it seems difficult to compensate the weight of intentions and strategic objectives of the firm (sometimes embedded in its culture, but also the functioning of the commissioned analysts). We can observe such a collective bias today, for instance in many foresighters' preferential attitude towards such dynamics as Open Source, Web 2.0, or wireless computing expansion, in which we can hardly distinguish the results of analysis from optimistic profiling of specific socio-technological developments. However, this type of rethoric is also converging with the expectations formulated by political leaders – who commission the analysts – regarding these particular Information Society dynamics.

Another problem is that most weak signals of 'important changes to come' do not necessarily appear as simple forms such as words, sentences, images or events, but rather as bundles of concepts and processes, systemic transitions, hybrid patterns of shifting influences, cultural change dynamics, etc. In others words, the most challenging changes we are confronted with involve complex processes with no easy clue to deciphering them. If we take as the referential unit not the firm but larger organisational or societal settings, for instance, global economics and politics enter the picture. A firm's strategic problem may be related to the rise of China and India as competitors in its traditional market, or be linked to emerging environmental taxes that have little to do with the firm or its specific domain of activity. How about larger issues such as international cybercrime or climate change? There are grounds to think that the Ansoffian scheme, with its filters and gradual response approach is just too simplistic to really deal with such complex and open-ended types of changes. It is particularly acute with highly disruptive innovations or transformations of the environment. We are here confronted by large scale but serendipitous changes which go beyond, both in terms of magnitude and depth, what Bower and Christensen (1995) mean by 'disruptiveness' (i.e. a form of radical innovation, process or business model change involving new products and new markets); disruptiveness, in the broader sense often implies a major deviation from current mindset or even a real paradigmatic shift, not just within one firm or sector of activity, but in society as a whole.

4. The notions of framing and meta-framing

4.1. *Establishing the condition for a reflexive distance*

As we have suggested in the previous section, one of the ways to reduce unpleasant aspects of the paradoxes and shortcomings identified so far is to consider weak signals, whatever they may be, as 'candidates' (or in other words, as hypotheses). The possible evolution of this status resides not only upon external forms of signalling, who would confirm its first evidence, and help build a case for stronger signals, but also in the possibility for the analyst (and perhaps his constituency) of: (1) framing some candidates so as to make sense of them in terms of weak signals (this is something one does inadvertently, most of the time) and (2) taking some reflexive steps to analyse the underlying principles and implications or his particular framing options, what we will call, from here on, 'meta-framing'. This does not solve entirely the series of problems evoked so far, but tends to transform paradoxical issues into reasonably working perspectives.

This claim becomes notably effective when it comes to tracing and tracking 'candidate weak signals' in time and space and evaluating whether they might have already become 'strong' (signals) or even more comminatory injunctions. In this approach, there is barely any possibility for a weak signal to exist outside of some vision- or scenario-building effort, hopefully involving the analysis of its underpinning rationale. In such a weak signal, 'construction' process, the implicit choices we make and decisions we come to, including all the embedded features and parameters which make them sustainable in time, constitute precisely what we call a framing operation. It means a certain way to select features, shape them into a consistent set of references or perspective, supported by more or less explicit boundaries, time considerations or contextual sense-making links and analysing at some point the assumptions underpinning our early detection activity (see for that Adam 2006). This set of characteristics, most often implied in the identification and interpretation of weak signals, once established as a domain of analysis, allows for the status of weak signals to become an important part of their sense-making potential. This is the substance of what we call meta-framing.

In reality, we do some form of framing activity all the time – in everyday life, in science, in managing organisations, in publicity and in political debate. Business cycles are also types of framing configurations, system modelling too, just as are Ansoff's filters. We are framing our world as a normal form of apprehending our environment and construct some mastery over it as a routine means of survival. We simply do not assess the implications of what we do. We are thus quite self-referential in our contextual interpretations and tend to naturalise what we observe, neglecting (1) the biases that we may convey, often not being aware of them, and (2), for our own convenience or/and by limitation of ours, the fact that we are constantly part of a multiple reality (others may make different selections and interpretations). If we manage to identify, in addition, the assumptions and conditions that we co-produce within our framing performance (i.e. the meta-framing level), then we can get a new degree of freedom in selecting, interpreting and combining our (weak signal) evidence into future-making activity. This is why it is important to understand what this notion of framing really involves.

4.2. *Framing: the history behind the notion*

Traces of the 'framing' notion can be found in earlier works, but the researcher who has really taken this concept up and promoted it to full status on the social science scene is Gofmann (1974), in his work 'Frame analysis; an essay on the organization of experience'.

Frame and frameworks are said to be more or less synonyms to context, background, setting, situations, etc., but as Gofmann goes on, it is clear that some specific attributes are at stake when using the particular notions of 'frames' and 'framing'. In particular, frames are supported and put in perspective in a quite differentiated manner by a whole series of other concepts like keying, events, tracking, fabrication, roles, rules, strips, etc.

A frame is a 'schemata' of interpretation of experience, allowing individuals to recognise events and build meaningful perception of what they capture (Gofmann 1974, 21). Framing brings together the activities of locating, perceiving, relating events, scenes or processes, as known patterns of meaning or 'labelling'. It always involves, even if unnoticed, some level of organisation (a perspective, an approach, or even a more strongly structured configuration activity). 'Framing however, organizes more than meaning, it also organizes involvement' (1974, 345), and there is no need to insist that this dimension is of great interest for management as well as policy-making. In all cases, framing is a way to sort out some forms of meaning, out of multiple realities, with more or less flexible boundaries, anchoring features and communicational implications. Finally, let us say that framing mechanisms are numerous: transformation, fabrication, connection, disconnection, confusion, etc. – it can generate false positives or false negatives (alerting, for instance, is a sort of framing). It is not surprising therefore, if a wide array of disciplines have, after Gofmann, taken up the notion of framing, including in the area of strategic management (see for that Kaplan 2004).

The link between this framing activity and weak signals is that most of weak signal identification and processing resembles a framing treatment allowing us to: (1) come up with options, (2) explore and test most promising meanings and implications, (3) elaborate responses to cope with strategic pressure.

In the case of a firm facing the kind of changes likely to take place within its specific environment, the tactics of top managers may be to play with Ansoffian schemes to exert a certain pressure and attempt to change the environment rather than to change themselves, or to go for minor adaptations, aligned to their competitors' moves. In this narrow type of context, the risk is limited for problems obviously not linked to global issues (although it is always difficult to establish, even afterwards, to what the firm' success or failure is really due). Finally, let us not exclude the possibility that some weak signal careers follow the Ansoffian scheme, but again, provided that some framing makes it consistent and feasible, and are captured by the organisation. Any major shift, on the contrary, will make the Ansoffian scheme less than effective in addressing emerging issues, as most of the indications are related to complex and broader types of changes and probably requiring important paradigmatic realignment in the short or medium term. In this case, framing is not sufficient and a true reflexive step is needed to diminish risks of false positives and negatives.

Let us finally insist on the fact that framing is, *per se*, non-reflective, but can be deconstructed as a inherent component of knowledge and communication processes which can be examined and even redesigned to incorporate reflective (or meta-level) features.

4.3. *From framing to meta-framing*

By introducing the necessary reflexivity to the possibility of self-questioning, we can transform framing into a more substantial concept: meta-framing. What do we really do when we 'frame'?

(1) We identify a topical context as reference.
(2) We suggest boundaries or contextual clues for validity or scope of usage (in time, space and societal context).
(3) We produce more or less explicit assumptions (actually, they are most of the time implicit).
(4) We build meaning, taking up a piece of information and configuring it as a situated form of knowledge (most of the time tacitly), with a particular perspective or through a particular form of communication.
(5) For complex arrangements (and their understanding), we are also capable of articulating several frames with one another.

All these dimensions, together – and provided that it is reflected upon – define the level we call 'meta-framing', which involves the capacity to establish as legitimate analytical objects (i.e. part of the methodology) these different shaping operations. The key to entering the level of reflexivity which can really make a critical difference in weak signal handling, however, is most probably number (3): making our assumptions as explicit as possible, and part of the weak signal identification process itself (i.e. taking into account the different usages we have of weak signals, according to our diverse roles and contextual interests).

Reflexivity, in that sense, means being capable of not only reflecting upon the 'what' (what is important, might be impacting), but also upon the 'how' and the 'why' (how and why we have come to select and interpret some events a certain way). For a more thorough linkage of reflexivity to futures studies, see Fuller, De Smedt, and Rothman (2006).

5. Managing change, coping with uncertainties: a pragmatic approach

Meta-framing may seem to be a universal solution, but it has two shortcomings: (1) although a meta-framing activity may help limit self-induced biases, there is no guarantee of identifying all the problems and clearing the situation being analysed, as if we could reach a sort of objective level of observation; (2) engaging in a meta-framing activity regarding uncertainties and strategic questions is quite time-consuming and cannot reasonable be envisaged all the time. A subsidiary question, therefore, is when to 'go for it', or when to rely upon more fluid and straightforward spotting and then processing of 'weak signals', in the Ansoffian style, but at the risk of missing crucial features.

A pragmatic approach has to be devised, suggesting that environmental scanning, along with weak signal analysis and management, as Ansoff and continuators have outlined it, should constitute a 'business as usual' basis, but that periodically, the organisation should be more explorative than just reactive, particularly about its own mindset and vision.

For business as usual, the tips provided by Mendonça, Steinmueller, and Hiltunen (see end of Section 2), cover a wide range of options from typical organisation's attention to modifications of their environment to more atypical actors' behaviour, and also bypassing the generation of wild cards. However, as we have stated that we cannot rely on our capacity to pre-evaluate and discriminate certain from uncertain, and impactless from impactful, weak signals, we must introduce other ways of dealing with uncertainties and emerging futures. In order to get a better chance to pre-evaluate the quality detection of trends as well as both false negatives or false positives, some level of framing and meta-framing analysis is suitable. It prevents us from blindly applying Ansoff filters, and so be capable of better differentiating what belongs on the one hand, to a changing reality to be captured in terms of weak signals, and on the other hand, to intentions and wishes interfering with the proper analytical capability of each of the above-mentioned stakeholders. This level of investment has to be considered for, but not restricted to, crisis-types of situations and, on the contrary, deployed according to several forms of organisational knowledge management and anticipatory moves. If the first activity consists of managing candidate weak signals to the point where a decision can be taken, the second category involves a more open and non-linear form of activity, in which what is managed, in fact, is our framing capacity.

In that perspective, each one of the following suggestions conveys its own lot of epistemic issues and new avenues for new in-depth research, also involving challenging fieldwork. In particular, there is a need for clarifying the respective roles of the various concepts through which the 'early' detection claim has been made in futures studies and strategic management. The weak signal reference was chosen here to keep it simple, but early warnings or near equivalents in other languages, as well as other metaphors (seeds, germs, etc.), and finally the rather inconclusive debates on wild cards, would help deepen the reflection initiated in this paper. At this level, however, we would like to limit ourselves to proposing the following options as a creative and complementary perspective to traditional weak signal theory in futures studies.

- The analyst commissioned by an organisation to identify and interpret weak signals in their early stage should beware of easy picks (the odd, the new and their linear projection into the future) and check for real system dynamics, often multiple and competing; instead of just sequential facts and gradual build-ups.
- This precaution also means that the weak signals of candidates should be routinely stored for further processing, whenever possible, and to engage in various concurrent weak signal analytical activity.

- When the moment arrives to start an open-ended analysis (not envisaged only as a reaction to external pressure), a pragmatic approach should prevail, balancing the most economic solutions, including quick methodological blends, with more explorative and reflexive forms of strategic analysis; as a general principle, short-term threats and decision-making habits (or power relation patterns) should not hinder the need for more in-depth periodic questioning.
- One interesting way of limiting most obvious biases is to start exploratory futures exercises with a backcasting approach rather than forecasting (Robinson 1982), linking weak signals with potential scenarios only as a concluding reflection rather than easy picks pushed into possible unnoticed preferences linked with forward thinking.
- In the case of the forward thinking approach to weak signal treatment, one way of limiting self-induced framing is to link weak signal picks to several cones of possibilities (the options that lie between two extremes) and not just one, thus making the Ansoffian process address questions of different nature, three or four clearly distinct questions defining a level of complexity usually beyond the analyst's control.
- Far from envisaging weak signal analysis as a one-size-fits-all approach, levels and scope of changes that an organisation has to cope with should influence the pragmatic philosophy suggested in this section. Responding to a competitor's move into the market, anticipating next summer's fashion is not an early detection problem of the same kind as dealing with the world financial crisis, the evolution of China's influence on mass-markets, or the impact of the first really severe oil shortage in the investment of energy-sensitive organisations or sectors. Adaptation to the scope and speed of the change envisaged should help match the different levels of possible uncertainties and surprises. Minimally, this means that the firm, as a strategic actor, has to envisage as obvious the evolution of its market and regulatory environment (the Porter-like approach to competitive analysis), but also the dynamics of more global pressures, constantly changing the landscape in which one evolves; hence the need to reflect on assumptions, models and paradigms.
- Uncertainties will remain, and instead of reacting to weak signals, the firm should also envisage no return points or tipping points for ineluctable decisions and actions; in an Ansoffian perspective, this means the *ex-ante* evaluation of when the firm considers that a weak signal will become a strong signal, and therefore force a response.

In all cases, what should prevail, as a generic way to limit one's own biases in the identification and treatment of weak signals, is giving constant attention to diversity (of the environment, the dynamics likely to take place, the interpretations of the same event or issue, etc.), converging in this sense with directions already evoked and suggested by Mendonça, Hiltunen and Steinmueller at the end of Section 2.

6. Conclusion

In this paper, we plead for a constructive re-visit of the very useful concept of 'weak signal'. We have examined the strengths, but also some deficits of the mainstream approach as outlined by Ansoff and continuators. We have emphasised that the main deficit of the work carried out so far in this now well-established perspective lies in the lack analytical reflexivity, a weakness regarding the analyst's own assumptions, boundaries, mindset, expectations and their overall impact in weak signal analysis itself. We drew upon the existing concept of framing, developed and proposed the more provocative idea of meta-framing as a stimulating area for better understanding such biased inflections of our own picks and further treatment of weak signals. We finally envisaged several ways of translating all these contributions into open and constructive managerial terms.

We are aware that going against the mainstream is not easy (dominant paradigms know how to strike back!) and that most of our proposals also have some level of paradox, including what we have criticised for the Ansoffian tradition. This is why we welcome – and consider an urgent need – any new research in that direction, not just reaffirming and enhancing Ansoff basics, but truly demonstrating that a re-enforced weak signal theory can help us cope with real challenges, not only for the firm, but also for society as a whole. The 'early' detection claim is of course the main problem tackled in this paper, but beyond that generic question there are more pragmatic issues to deal with, such as those presented in Section 5, but more broadly all discontinuities and radical shifts from current trends, which are key concerns for innovation but also risk management at both the micro-level (the firm, its neighbouring environment) and macro-level (changes impacting the global market and beyond, possibly, the whole of society).

Notes

1. See in particular Ansoff (1975, 1982, 1984).
2. For a systematic attempt at clarifying the distinction between weak signals and wild cards, see Mendonça et al. (2004) and Hiltunen (2006).
3. See www.fountainpark.com/signals_solutions/
4. Hiltunen (2007) is also engaged in a more creative approach to weak signals, dedicated interact with a variety of actors, using the support of images as windows on possible futures (http://www.flickr.com/groups/weaksignals).
5. On the constructivist side, it is worth mentioning the pioneering attempt of Seidl (2004).
6. A good example of that is Webb (1987) who looks for difficulties of identifying weak signals mostly in problems related with the signalling source.

References

Adam, B. 2006. Has the future already happened? Paper presented at the International Conference, 'Future matters: futures known, created and minded', 4–6 September 2006, Cardiff University,

Ansoff, I. 1975. Managing strategic surprise by response to weak signals. *California Management Review* 18, no. 2: 21–33.

Ansoff, I. 1982. Strategic response in turbulent environments. Working Paper No. 82-35, European Institute for Advanced Studies in Management, August 1982.

Ansoff, I. 1984. *Implanting strategic management*. New York: Prentice-Hall.

Bower, J., and C. Christensen. 1995. Disruptive technologies: catching the wave. *Harvard Business Review*, 73, no. 1: 43–53.

Choo, C.W. 2007. Information life cycle of emerging issues. http://choo.fis.utoronto.ca/ncb/es/ESinfoLC.html.

Coffman, B. 1997. Weak signal research. MG Taylor Corporation, Louisville KY, USA. http://www.mgtaylor.com/mgtaylor/jotm/winter97/jotmwi97.htm.

Day, G. S., and P. J.H. Schoemaker. 2006. *Peripheral vision: detecting the weak signals that will make or break your company*. Cambridge, MA: Harvard Business School Press.

Festinger, L. 1957. *A theory of cognitive dissonance*. Evanston, IL: Row, Peterson.
Freeman, O., and H. Pattinson. 2007. Scenarios for the client in scenario planning: six walks in the scenario woods. Paper presented at the Third International Conference on Organizational Foresight 'Learning the future faster', 16–18 August 2007, The Strathclyde Business School, University of Strathclyde, Glasgow.
Fuller, T., P. de Smedt, and D. Rothman. 2006. Advancing foresight methodology through networked conversations. Paper presented at the Second International Seminar on 'Future-Oriented Technology Analysis (FTA)', September 2006, Sevilla.
Gilad, B. 2003. *Early warning: early warning: using competitive intelligence to anticipate market shifts, control risk, and create powerful strategies*. New York: AMACOM.
Gofmann, E. 1974. *Frame analysis: an essay on the organization of experience*. London: Harper and Row.
Gruber, M., B. Kolpatzik, J. Schönhut, and C. Venter. 2003. Die Rolle des Corporate Foresight im Innovations-process. *Zeitschrift Führung und Organisation* 5: 285–290.
Hiltunen, E. 2006. Was it a wild card or just our blindness to gradual change. *Journal of Futures Studies* 11, no. 2: 61–74.
Hiltunen, E. 2007. The futures window – a medium for presenting visual weak signals to trigger employees' futures thinking in organizations. HSE Publications, Working paper, w-423.
Hiltunen, E. 2008. The future sign and its three dimensions. *Futures* 40: 247–260.
Holling, C.S. 2004. From complex regions to complex worlds. http://www.cof.orst.edu/starkerlectures/transcripts/2003/holling.pdf.
Ilmola, L., and O. Kuusi. 2006. Filters of weak signals hinder foresight: monitoring weak signals efficiently in corporate decision-making. *Futures* 38, no. 8: 908–924.
Kamppinen, M., O. Kuusi, and S. Söderlund. 2002. Tulevaisuudentutkimus, Perusteet ja sovellukset (in Finnish). Helsinki: Suomalaisen Kirjallisuuden Seura.
Kaplan, S. 2004. Framing contests: micro mechanisms of strategy-making in the face of technical change. Working paper. http://www.london.edu/assets/documents/PDF/Paper_Kaplan_1.pdf.
Könnölä, T., V. Brummer, and A. Salo. 2007. Diversity in foresight: insights from the fostering of innovation ideas. *Technological Forecasting & Social Change* 74: 608–626.
Kuosa, T. 2005. Heikko signaali vai merkityksetön kohina: pattern management – ontologisesti uusi lähestymistapa heikkojen signaalien tarkasteluun ja tulkintaan (A weak signal or meaningless noise: pattern management – an ontologically new approach to examining and interpreting weak signals). *Futura* 19, no. 4: 115–120.
Levinthal, D., and J. March. 1993. The myopia of learning. *Strategic Management Journal* 14: 95–112.
Mannermaa, M. 2004. Heikoista signaaleista vahva tulevaisuus (Creating strong future with the help of weak signals). Porvoo: WSOY.
Mendonça, S., M. P. Cunha, J. Kaivo-oja, and F. Ruff. 2004. Wild cards, weak signals and organizational improvisation. *Futures* 36: 201–218.
Morrison, J. and I. Wilson. 1986. The strategic management response to the challenge of global change. In *Future vision, ideas, insights, and strategies*, ed. H. Didsbury. Bethesda, MD: The World Future Society.
Morel, C. 2002. *Les décisions absurdes*. Paris: Gallimard.
Petersen, J. L. 1999. *Out of the blue – how to anticipate big future surprises*. Lanham, MD: Madison Books.
Robinson, J. 1982. Energy backcasting: a proposed method of policy analysis. *Energy Policy* 70, no. 4: 377–344
Rossel, P. 2007. Meta-framing: the art of putting weak signals in perspective. Paper presented at the COST A22 Final Conference 'From Oracle to Dialogue, Exploring New Ways to Explore the Future', 9–11 July 2007, Athens.
Seidl, D. 2004. The concept of weak signals revisited: a re-description from a constructivist perspective. In *Managing the future: foresight in the knowledge economy*, ed. H. Tsoukas and J. Sheperd, 151–168. Oxford: Blackwell.
Snowden, D. 2002. Complex acts of knowing: paradox and descriptive self-awareness. *Journal of Knowledge Management* 6, no. 2: 100–111.
Snowden, D. 2005. Strategy in the context of uncertainty. *Handbook of Business Strategy Journal* 6, no. 1: 47–54.
Steinmueller, K. 2004. Wild cards – using wild cards in influencing change. http://www.thegff.com/futurescene2004/Presso04/05.%20K.Steinmuller.Pdf.
Steinmueller, K. 2006. Wild cards and trends. Zpunkt, the Foresight Company. http://www.interact-eu.net/download/application/pdf/1016804
Webb, J. R. 1987. An evaluation of Igor Ansoff's theory of weak signal management by means of an investigation and forecast of future developments in the ophthalmic laser environment. Doctoral thesis, University of Strathclyde, Department of Marketing, Glasgow.

Exploring the weak signals of starts-ups as a folksonomic system

Olivier Glassey

Observatory of Science, Policy and Society, University of Lausanne, Switzerland

This article looks at the opportunities for innovation and entrepreneurship provided by the swift technological change in the field of information technologies: the so-called Web 2.0. It focuses on how it could be possible to analyse emerging trends from actor's strategies toward the future. The main research question is to understand how early signals of new trends appear in a field which is not formally stabilised. We looked at the spontaneously produced categorisations (folksonomy) of 250 start-ups. This corpus is analysed in order to explore how these descriptions identify key future evolutions and how these categorisations of future trends change over time. Our central hypothesis is to consider that early elements of future changes (weak signals) are not given once and for all, but exist within perpetual shifting processes and could be tracked with folksonomic maps. The discussion of the results will suggest that weak signals should not be conceptualised as static entities blessed with some strange power to shape the future but as a set of components, whose main features are their abilities to be constantly rearranged among themselves by the involved actors.

1. Looking for seeds of change

The main intention of this paper is to tackle the question of the origins of future changes. This question has a long lasting tradition among researchers in foresight (Ansoff 1975, 1985; Coffman 1997; Blanco and Lesca 1997; Mendonça et al. 2004; Saul 2006). Those approaches used a large array of metaphors to embody the idea of 'elements announcing further (r)evolution' (weak signals, early warning signals, wild cards, seeds of change, etc.) While sharing a common interest these approaches do not propose a common ground to define what the basic components of potential changes are. In order to define our own perspective we decided to take from this rich diversity the idea that it is hardly possible to identify, as specific objects, something that could be described as 'the seeds of change'. We decided to start back from a general level and consider how classical approaches to this question of early warning could be categorised schematically in two main trends.

The first trend consists of the capitalisation on the former experiences by mining a vast corpus of cases provided by history, in general, and history of innovation and technology, in particular (Martino 1983). Sociology, history of technology and especially constructivist approaches developed since the late 1970s,[1] offer a great number of studies and conceptual or methodological tools, which were helpful in understanding the trajectory of innovation. Traditionally such a methodology gathered valuable insight to sociological, economic and historical perspectives on innovation. These analyses which, in their majority, overcome the double hurdle of social or technological determinism (Roe Smith and Marx 1994), allowed us to establish how innovations processes were the results of iterative interactions within an heterogeneous network of 'actors' (ranging from engineers, investors, politicians to end-users) and technologies. These studies point out that the outcome of these negotiation processes influence the fate of an innovation. They also help to clarify the dynamics of innovation and stress how, negotiation on technical (norms, formats) and social aspects (intended users), could have a shaping power in the long term on the innovation trajectory. In that way these studies could help us to identify elements that are candidates to play the part of seeds of change. However, *ex post* analysis, when used in a foresight perspective, is exposed to potential rationalisations, which are often embedded in the narratives supported by these approaches (Hennion 2003). Along with Schnaars (1989), we could argue that the detection of past seeds of change is a poor predictor for present and future situations because it is built upon the influential and non-trivial knowledge of who 'won' (i.e. what is the fate of the past trends). The use of a symmetrical analysis of successes and failures (Bijker, Hughes, and Pinch 1987; Bijker and Law 1992) could reduce this tendency.[2] However the majority of research has trouble putting the same emphasis, on the one hand, on success stories and, on the other hand, on apparently meaningless failures. What is more, the symmetry principle is seldom fully implemented. Indeed, the major obstacle that allows the transfer of experience gained in the field of sociology and history of technology to the field of future studies is related to the tendency to undermine a key ingredient of any foresight analysis: uncertainty. Thus, it is difficult for them to reproduce *ex post* the situation of uncertainty, in which the innovation trajectory existed. This uncertainty could be considered as a defining component to understand actors' behaviours through the options and the strategies they choose.

The opposite approach relies on an *a priori* identification of potential seeds of change. As stated before, this second category of approach is popular among the specialists in the field of future studies. It is at the core of numerous forecasting (Schwarz, Svedin, and Wittrock 1982) or backcasting (Robinson 1990) methods designed to define how to build appropriate scenarios of potential futures. Beyond specific methodological traditions, those scenarios are mostly used in groups. The presentation of future possible trends is translated in the design of a set of scenarios that highlight different options for the future. In order to be understandable by their users, those scenarios must inevitably be somewhat contrasted. These links among the scenarios (the variable identified as valid to characterise different scenario options) define the frame and, at the same time, suggest a definition of the elements considered as relevant in structuring the future. In this perspective, the seeds of change are somehow imbedded at the core of the production modes of scenarios. This way to produce potential seeds raises yet another problem: the definition of the object to observe. It is difficult (perhaps impossible) to conceptualise an approach that anticipates any kind of seeds of change without, simultaneously, inducing a theory about the ontology of these elements. The construction of such scenarios is often framed trough the preoccupation of designers and thus by their definition of what is pertinent to observe.

In order to minimise this double pitfall of *a priori* projections and *a posteriori* rationalisation, this paper would like to suggest an alternative strategy. Our starting assumption postulates that, strictly speaking, there is no such thing as an ontological category of 'seeds of change'. In other words, we propose the idea that no isolated element can fulfil, by itself, the role of influencing the future. On the basis of our hypothesis we need to adapt the focus of our investigation. Instead of seeking for such 'seeds', we should rather try to identify how heterogeneous elements are combined together to frame possible evolution. Metaphorically, one could say that we should not try to define a methodology to build a new set of radar towers for detection of weak signals, but rather recreate the condition of a micro universe in which actual evolution of trends aimed at the future could be tracked and understood in their contexts.

Our principal hypothesis is to consider that the emergence of new changes cannot be done independently of the way people involved in these processes try to represent and understand them, both at individual and at group level. In other words, the socialisation of representations about the future, as they are embedded in the strategies of the actors, constitutes a privileged place for a better understanding of the emerging process of new tendencies.

The creation and organisation of such observation requires us to take some precautions. Among these precautions we must pay specific attention to avoid creating unnecessary determinism where uncertainty exists. We have tried as much as possible not to provide any *a priori* definition of potential candidates that could belong to the whole of elements that could influence the future. As the aim of our research is not to seek reliable forecasts (for which we would have no bases to assess their accuracy) it appeared to be more interesting, because it is more open, to delegate the responsibility for the definition of such relevant elements to the actors themselves. Because our goal is more focused on collecting the diversity and the richness of the representations, through which these actors try to grasp the future, we opted to utilise actor-made folksonomies as an object of study.

2. A folksonomics approach

Folksonomy belongs to the family of what could be described as 'distributed classifications' (Shirky 2005). Categories in a folksonomic process are not produced by a group of experts which control the whole classifying structure and who validate a common thesaurus to use. Folk classification or ethno-classification (Berlin 2002) emerge directly from users' contributions (Merholz 2004). This means that in folksonomies users are free to label categories according to their own preferences; for example, the millions of photographs or videos which are available online on sites such as Flickr or Youtube and are described by a series of labels which characterise them in a non-systematic way. These labels, known under the name of tags, are in these cases proposed by the people who provide the content. To establish these tags, users are 'free' as they can tap into a natural thesaurus: their own imagination (or other users' imaginations). The core principle of these bottom–up classification processes consists in offering to users pragmatic means of contents navigation that are close to their concerns and their own spontaneous classifying practices (Kroski 2005). As stated by one of the sites promoting tags use, the advantage of tags lies in the fact that they are '... a little bit like keywords, but they're chosen by you, and they do not form a hierarchy. You can assign as many tags to a bookmark as you like and rename or delete the tags later. So, tagging can be a lot easier and more flexible than fitting your information into preconceived categories or folders.'[3]

The analyses conducted on the nature of these user generated thesauruses underline the heterogeneity of the utilised notions (Guy and Tonkin 2006). They could cover a broad array constituted by subjective concepts ('super', 'cool', 'interesting'), phonetic diminutives, misspelled words and even by multilingual linguistic innovations of Creole type (Golder and Huberman 2006). These elements form a kind of 'semantic soup' that has no hierarchy. This lack of structure is obviously one of the main weaknesses of the folksonomic systems, which often fail to generate any kind of guiding principles for user orientation within the classification items. This limit concerning a structuring framework is particularly sensitive at the level of synonyms' management. Indeed, in open folksonomies there are no tools to manage semantic ambiguities (Kroski 2005). Another issue is the lack of lasting consistency because of the unending evolution of tags that are in a state of constant (re)edition and thus make it impossible to guarantee the meaning of used tags.

Considering these limits, it is legitimate to wonder whether folksonomies are more a factor of entropy increase than a means to reduce it (Glassey 2007). It could also be argued that its lack of formal structure would prevent any valuable data mining from such corpus. The examination of the perceived benefits of these classification modes, however, could bring some nuances to this early assessment.

In fact, some of folksonomies' identified weaknesses are also among the key reasons of their popularity. For example, the absence of hierarchy is undoubtedly problematic but it also lowers the barriers of access to the creation and use of classification. To make a query within a folksonomy, the user does not have to go through the process of acquiring a mental map of the hierarchal structure of some classification tree. He or she does not need to think about the semantic accuracy and deal with possible synonyms to have access to results that could turn out as meaningful. The simplicity of free typing and loose associations is a lot easier than making a decision about the degree of match (Mathes 2004). In practice, classical taxonomies are very limited when it comes to dealing with a very large and fast-evolving content pool. In these cases, which cover many basic every-day life internet uses, the obligation to learn a complex, hierarchical controlled vocabulary would just be too costly for many users in terms of time and cognitive effort (Mathes 2004). In stark contrast, within large tag pools, basically any word (even misspelled or contracted such as in SMS language) will bring some results more or less loosely related to the intended query. To narrow down uncertainties linked to folksonomy's messy classification, people often use a combination of several tags which could lead to more precise results. Tag combinations are powerful and flexible tools that allow building original hybrid categories, which are directly inspired by the contributors' representations. Practitioners involved in knowledge management issues have already identified folksonomy as a potential field to enhance existing knowledge sharing platform (Lund et al. 2005). In contrast with the heavy management required to build metadata platforms and semantic web initiatives, folksonomy offers a rough but easier way to implement navigation tools.[4]

The use of the tag-induced folksonomies does not go without raising problems, in particular, because of its lack of precision, among others (Guy and Tonkin 2005). Within the framework of our study, these inaccuracies, mistakes or ambiguities do not constitute a setback but, on the contrary, they are a means of revealing the complexity, the creativity and the diversity of the points of view regarding the description of a common field of interest. We think that this somewhat chaotic diversity also belongs to the above-mentioned principle of uncertainty that, in the practice of foresight studies, should not only be understood as 'the uncertainty about the future' but also about the present. In the context of everyday life, uncertainty could also be related to diverging interpretation, approximations, groping and hesitations of people and this is also true when they participate in presenting future activities.

3. Method

Since 2005, the 'Web 2.0' as quoted by Tim O'Reilly (2005), has become a trendy notion with fuzzy boundaries, synonymous for many with a paradigm shift of the internet evolution. Nonetheless, its definition remains problematic as it covers a vast array of heterogeneous, technical (Ajax, Soap, API Services, mash-ups, etc.), social (folksonomy, user as contributors, social mimetic services, etc.) and organisational processes (mass service of micromarkets, radical decentralisation, etc.). This notion is also used as a generic means to describe the evolution of the internet functionalities, services and use at technical, economical and social levels. Furthermore, the success of the term was such that the '2.0' tag started to proliferate and is now more and more used to describe innovation as a whole, even when it has limited connections with the web. The lack of a stabilised simple definition of what actually is 'Web 2.0', does not prevent this concept from being extraordinarily popular among the people involved in information technology. However, some observers do not hesitate to question whether this concept has any substance and specificity, if it is only a marketing label aimed at giving a new youth to technologies and processes that have already existed for a long time, but suffered from the burst of the internet investment bubble (Travers 2007). Beyond such controversies, this notion interested us because the proliferation of multiple uses of the term 'Web 2.0' expresses a plurality of visions of the future of information technology, which is what we were looking for.

The interest in the question of Web 2.0's definition does not only refer to theoretical stakes. As a conceptual frame it also operates as a tool, an enabling frame providing a specific context from which it is possible to build strategies and mobilise resources (Shuen 2008). For example, this definition of what covers Web 2.0 plays a big role for innovators as a mean to foster financial partners' interest or media and customers awareness about the idea of a new generation of services. As such, the label 'Web 2.0' operates in start-ups as a flag for rallying potential business allies. In this perspective the fast and somewhat blurry evolution of this field does not only constitute an obstacle (to scan the landscape of existing initiatives) but also an opportunity to propose their own vision of the future in order to present their intended activities as strategic part of it.

For a better understanding of these individual bottom–up future scenarios, we have done a systematic analysis of the description of the core business of over 250 self-proclaimed Web 2.0 start-ups. We elected to focus on start-ups because, as newcomers, they have to present and defend their activities as being at the core of significant future evolution, in order to attract further financing resources or customers, but before looking at the corpus of those descriptions and the hints they could provide about some kind of spontaneous and distributed foresight we had first to consider the nature of what we intended to track.

In order to find folksonomies related to our topic, we have gathered a corpus of how start-ups, which explicitly refer to this notion of 'Web 2.0', are characterised mostly by third party commentators. We built a database of folksonomic descriptions of start-ups from different online directories which provided such functionalities. Concretely, we have collected the terms used to describe the activities of start-ups in order to produce an open thesaurus organised only by the representations and the associations made by the contributors of these folksonomic classifications. We did not proceed to any selection according to the field in which these start-ups propose their services nor did we make any assessment of the viabilityof the projects or their stage of development (restricted beta, beta, running, etc.). The geographical locations of the initiatives were also not taken into account and we only kept as the only relevant element of classification the date when the folksonomic assessment of these companies was made.

Folksonomic descriptions were collected on various web sites that maintain a list of Web 2.0. starts-ups in order to widen the size and the diversity of data collection. The sites were pre-selected according to the three main criteria. The first one is the existence of folksonomic descriptions as some sites dealing with Web 2.0 start-up do not include a tagging function. The second is the consistency of the corpus as many websites were short-lived initiatives and did not provide enough material for the full selected time frame. The last criterion is the existence of a generalist corpus as a few directories are specialised in one subtype of starts-ups and thus provide a narrow vision of the field. The three selected websites[5] also offer a diversified folksonomic production modes and merge descriptions produced by the teams in charge of the sites, descriptions provided by the starts-ups themselves or descriptions generated by users who describe and assess the potential of the various projects.

Among the examined cases (500) in three main lists, we have selected a set of 250 companies according to the date of their folksonomic description. We have divided our corpus into two groups following a single rule: the first group is formed by the first 125 start-ups described after November 2005 and the second group is formed by the first 125 companies described after the beginning of 2007. The sample of cases was done in order to follow two main requirements. The first one was to obtain two groups of identical size, which allow for a comparison of the density and the diversity of proposed descriptions. The second requirement was to have several distinct groups that remain relatively close in time (the gap between the last description of the first group in June 2006 and the first description of the second group in January 2007 is less than seven months) in order to grasp the pace of the observed phenomenon.

On this basis we have built a chart of the terms used and how they are connected to each other (co-presence within the same description). We did not intend to produce an exact picture of the state of the Web 2.0, but only to grasp how what is projected to exist at a specific time is depicted and how it is changing over time.

4. Results

In our first analysis, the results show a strongly contrasted situation between the two groups of start-ups that we have selected. Whereas the number of studied cases is identical, the structure of the relations is definitely more packed for the most recent group.

The means of describing the potential and the possible applications of 'Web 2.0' has grown considerably in a few months. In the case of the start-ups described in 2005–2006 (Figure 1), the pattern is quite structured. Among the most important terms, i.e. those that obtain a central position within the cloud of quoted elements, we can point out: [1] communities; [2] social bookmarking; [3] collaboration: [4] data storage (download-hosting); [5] e-commerce; [6] search engine.

This structuring, which we cannot examine here in detail, is strongly built through a series of 'traditional' terms borrowed from 'classical' domains (e-commerce, virtual communities, etc.) that have existed for almost a decade. Whereas the concept of 'Web 2.0' was still young, the majority of the references related to an older model of using the internet. This is especially striking when we consider how many terms, which were to strongly organise the same field a couple of months later, are connected only to a subpart of the network. For example, 'share', which is among the most cited terms in 2007, is mainly related in 2005–2006 to 'photo sharing' within the group 'storage'.

Less than a year later the situation radically evolved and the diagram – obtained by exactly the same method – shows a quite different and more complex landscape of notions (Figure 2).

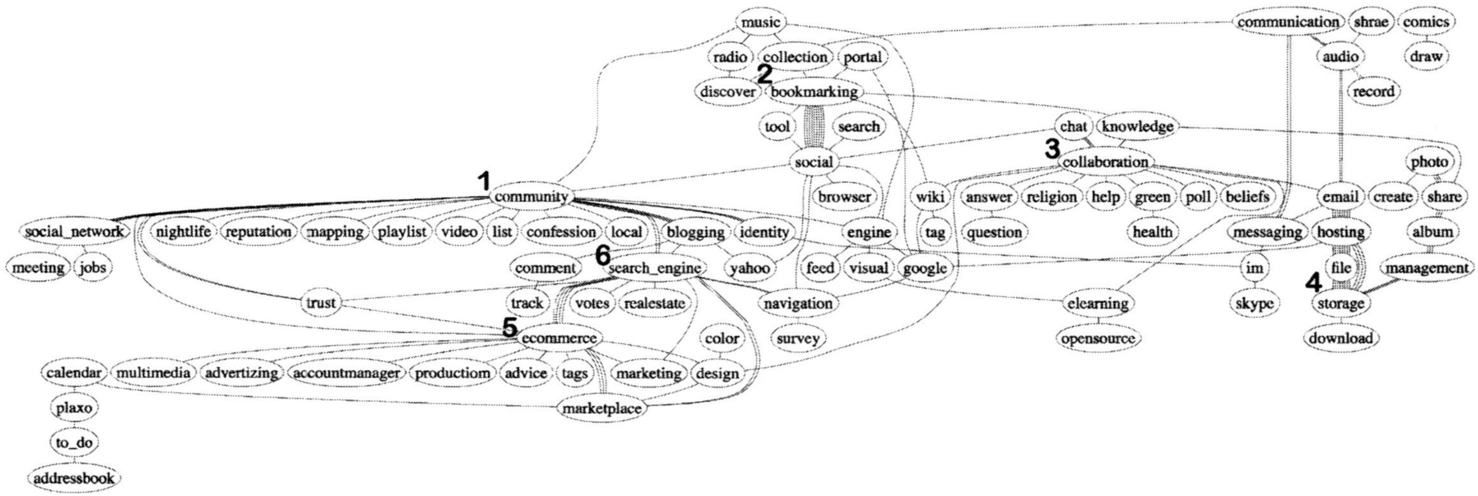

Figure 1. The map of folksonomic items used to describe 125 start-ups in 2005–2006.

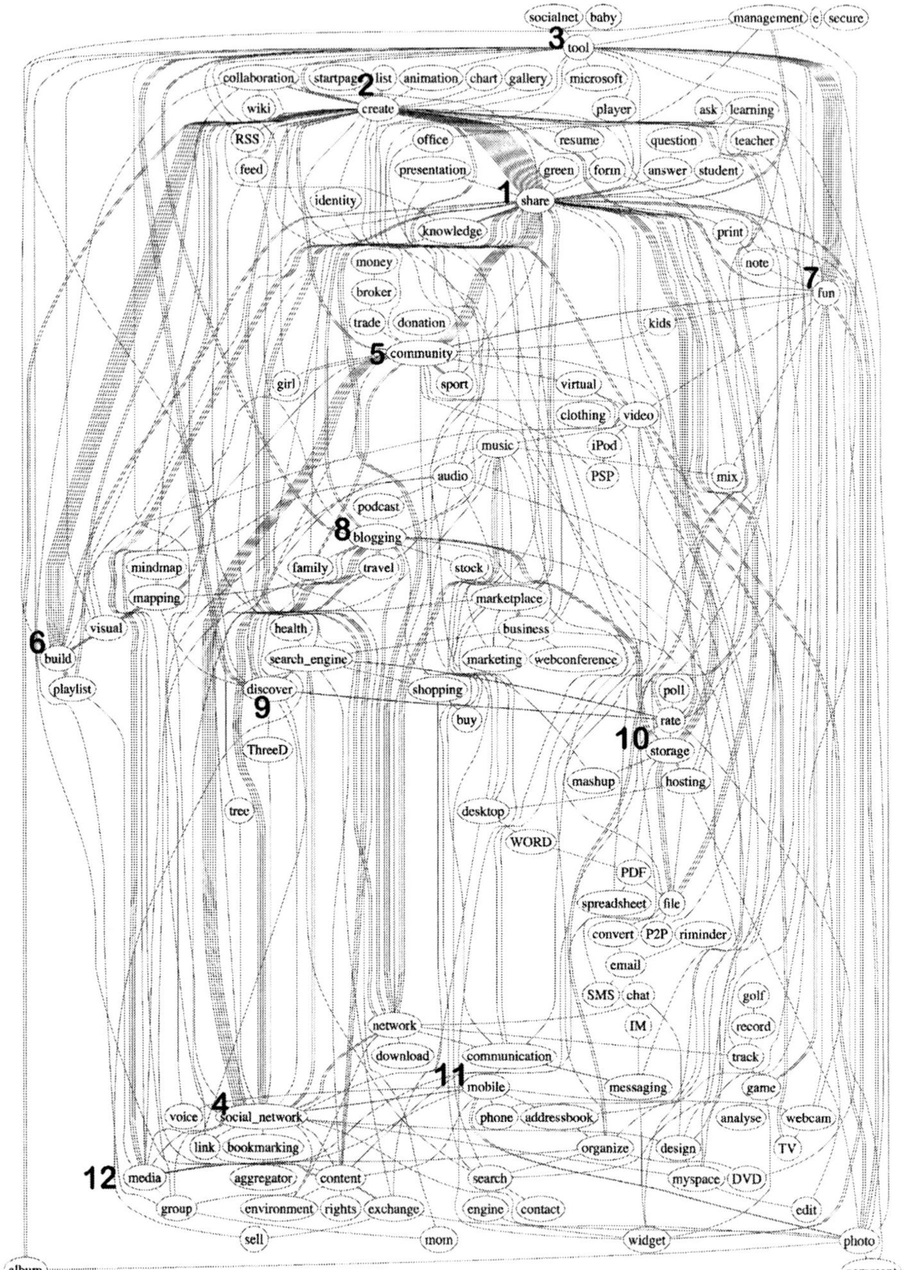

Figure 2. The map of folksonomic items used to descibe 125 start-ups after 2007.

The general structure is more difficult to outline as the amount of the relations between the descriptive elements has proliferated. It is however still possible to distinguish the appearance of new attractors. The terms occupying a central position are:

[1]	share	[7]	fun
[2]	create	[8]	blogging
[3]	tools	[9]	discover
[4]	social networks	[10]	storage
[5]	community	[11]	mobile
[6]	built	[12]	media

We have noticed that the majority of key terms used to describe the Web 2.0 start-ups were completely renewed. One can make the assumption that the movement observed between the two groups underlines the fact that a process of social learning and appropriation of the possibilities of these technologies is taking place. Looking at both maps gives the impression that only limited links exist between them. A closer analysis shows however that certain terms are actually transformed and have morphed into other terms. For example, the concept of *community* is present in the two maps, but, in the second one, some dimensions which were a subset related to *community* have become a specific trend (*social networks*). What we can observe is not only the disappearance of some elements as a valid means to describe on going evolution, but also the fragmentation, reconstruction and the strengthening of some notions through more specific descriptors.

Whereas the first group rested, essentially on categories and relations between categories clearly established in the former model, the second group suggests a dissimilar situation with the appearance of an original main cluster. This main cluster is built through strong connections between the terms 'share', 'create', 'tools' and 'build'. This new core could be interpreted as a major shift and we can make the hypothesis that this field is developing its own models of how to consider future innovation. The increased link density and the absence of semantic islands are also indicative of a resilient crossing between the various descriptive elements. Even if it were possible to distinguish some centrality, the interconnection of those elements is less univocal than in the earlier group. We would argue that, in the interval which separates the two groups, we witnessed the emergence of a platform made of a set of interrelated themes from which it is possible to envision the dynamics of a series of future developments.

As we have already mentioned, our goal in this paper is not to try and draw up an exhaustive range of possible future trends of Web 2.0. We are looking, more specifically, at how representations of the future evolve through our analysis of the folksonomics' categories applied to a series of start-ups. Beyond the subjective representations of each actor, we have obtained a glimpse of a broader landscape, more complex, and more chaotic, which is in perpetual reconfiguration. In a foresight perspective, this landscape is interesting because it represents the present state of a dynamic network which is constructed on how people try to address the future (in our case promote the activities of start-ups). In consequence, what is meaningful in these maps is not related to the presence of any specific elements but rather to the synergies that are built between various elements. These synergies, constructed and enacted by heterogeneous actors, define a perimeter from which future trends can spread. We believe that such maps could offer a fruitful background to observe how critical mass phenomenon processes (Ball 2004) or that crowds' wisdom as described by Surowiecki (2004) could play their part in the shaping of the future. Our approach which focuses on 'user driven' semantic offers an original perspective of the landscape of the future which innovation promoters build, together, through the framing of their own activities.

5. Conclusions

Beyond the search for hypothetical seeds of change, we were interested in a better understanding of how collective representations of the future could emerge through a multitude of prospective strategies. We have defended the idea that an approach in terms of folksonomic clusters could be adapted to accomplish such a task. The study presented in this paper is still in an exploratory stage and we are quite conscious of the limitations of the suggested approach which cannot claim to take into account (but could track) unanticipated shifts and disruptive innovations. Nonetheless, we think that this method could provide some clues toward a better understanding of the processes linked to the emergence of change within a community of specific actors. Its main feature is to offer a way to track the dynamics leading to future trends within a heterogeneous whole which composition and inter-connexions is unceasingly reconfigured by the actors themselves. This folksonomic approach is flexible and offers an alternative solution to draw up cartographies of fields which escape an exhaustive formal description. In other words, by delegating the definition of the observation framework to the actors themselves, the described approach does not only set these actors as the focus of prospective analysis but also set them in the middle of the actual definition of this future. From our point of view, this is just the first step of an overall research programme that aims at reflecting how the concept of the mutual shaping process, popularised in the STS field, can be interpreted in a foresight perspective in order to explore how the future is built by, but also builds, the present.

In our view, the agenda for further research in this field should simultaneously deal with methodological questions and operational issues regarding the strategic use of this type of data in actual practices.

At the first methodological level the aims will be to improve the production and the analysis of folksonomic maps by identifying more relevant corpuses and developing better tools for data processing. Our first results were presented in the form of two snapshots but it would be interesting to a track, in simulated real-time, the evolution of the observed fields. A better comprehension of folksonomy's dynamics will also necessarily pass by a qualitative analysis of their producers in order to understand how these actors describe emergent processes and how they adapt these descriptions according to the context evolution.

At the operational level further research should be conducted to assess how these maps can provide guidelines for action and function as collective watch systems. We have to keep in mind that folksonomies are not stand-alone products. The building of a folksonomic 'space' engenders the creation of a community of users which, in turn, offers many advantages not only to the users but also to the project promoters. Popular websites such as *Google, del.icio.us* or *Digg*, which work as scanners of Web content are partially built on a huge folksonomic thesaurus for whatever is considered by each user as 'interesting'. In the same philosophy but with the specific public of scientists as a targeted audience we find sites like *Connotea*, launched in 2006, which let the users store and arrange scientific article references with tags. Provided that it is a success (i.e. adopted by a significant number of scientists), such a service could give to their promoters an extraordinary insight into how scientists map their knowledge and build/update their reference lists and even how they work. The business of large scale folksonomy-based services is already up and running and needs to be explored. This exploration should lead to further scientific investigation such as, for example, how such private or public thesauruses could be used to identify fast evolving fields and work as early detection tools. It will also raise the question of what kind of knowledge is extracted from these massive folksonomic pools and, ultimately, how this knowledge could be harnessed to actual strategies.

Notes

1. The sociology of scientific knowledge (SSK), the social shaping of technology (SST) and the social construction of technology (SCOT).
2. These authors suggested an adaptation of the symmetry principle as defined by Bloor (1976) in the field of sociology of science: the demand that true and false beliefs (or, in the case of technology, both devices that work and those that fail) – should be analysed in the same terms.
3. http://del.icio.us/help/tags.
4. For an example of the use of folksonomy deployed within some large companies, see Fitzgerald (2006).
5. The online directories used to build our database that was providing tags for the description of the start-ups were: http://www.go2web20.net/, http://web2.0slides.com/, http://www.buzzshout.com/.

References

Ansoff, I.H. 1975. Managing strategic surprise by response to weak signals. *California Management Review* 18, no. 2: 21–33.

Ansoff, I.H. 1985. Conceptual underpinnings of systematic strategic management. *European Journal of Operational Research* 19: 2–19.

Ball, P., 2004. *Critical mass: how one thing leads to another*. London: Arrow Books.

Blanco, S., and H. Lesca. 1997. Environmental scanning: designing a collective learning process to track down weak signals. Paper presented at the Third American Conference on Information Systems (AIS) Conference, 15–17 August, Indianapolis, USA.

Bloor, D. 1976. *Knowledge and social imagery*. Chicago: University of Chicago Press.

Berlin, B. 2002. *Ethnobiological classification: principles of categorization of plants and animals in traditional societies*. Princeton, NJ: Princeton University Press.

Bijker, W. E., T.P. Hughes, and T.J. Pinch. 1987. *The social construction of technological systems: new directions in the sociology and the history of technology*. Cambridge, MA: MIT Press.

Bijker, W.E., and J. Law, eds. 1992. *Shaping technology/building society*. Cambridge, MA: MIT Press.

Coffman, B. 1997. Weak signal research, 15 January, MG Taylor Corporation. http://www.mgtaylor.com/mgtaylor/jotm/winter97/wsrintio.htm

Fitzgerald, M. 2006. Tagging tools offer powerful way to organize information. *CIO Magazine*, April. http://www.cio.com/article/19682/Tagging_Tools_Offer_Powerful_Way_to_Organize_Information

Glassey, O. 2007. When taxonomy meets folksonomy: towards hybrid classification of knowledge? Paper presented at the Proceedings of the ESSHRA conference 'Towards a Knowledge Society: Is Knowledge a Public Good? Dynamics of Knowledge Production and Distribution', 12–13 June 2007, Berne, Switzerland.

Golder, S., and B.A. Huberman. 2006. Usage patterns of collaborative tagging systems. *Journal of Information Science* 32, no. 2: 198–208.

Guy, M., and E. Tonkin. 2006. Folksonomies, tidying up tags? *D-Lib Magazine*. http://www.dlib.org/dlib/january06/guy/01guy.html

Hennion, A. 2003. L'innovation comme écriture de l'entreprise: récits d'innovation au sein d'une entreprise de services. In *Encyclopédie de l'innovation*, ed. P. Mustar and H. Penan, 131–152. Paris: Economica.

Kroski, E. 2005. The hive mind: folksonomies and user-based tagging. blog InfoTangle. http://infotangle.blogsome.com/2005/12/07/the-hive-mind-folksonomies-and-user-based-tagging/

Lund, B., T. Hammond, M. Flack, and T. Hannay. 2005. Social bookmarking tools (II). A case study – Connotea. *D-Lib Magazine*. http://www.dlib.org/dlib/april05/lund/04lund.html

Martino, J. 1983. *Technological forecasting for decision making*. New York: Elsevier Science.

Mathes, A. 2004. Folksonomies – cooperative classification and communication through shared metadata. Computer Mediated Communications, Graduate Shool of Library and Information Science University of Illinois Urbana-Champaign. http://www.adammathes.com/academic/computer-mediated-communication/folksonomies.html

Mendonça, S., M.P. e Cunha, J. Kaivo-oja, and F. Ruff. 2004.Wild cards, weak signals and organizational improvisation. *Futures* 36: 201–218.

Merholz, P. 2004. Metadata for the masses. Adaptive Path, San Francisco, CA. http://www.adaptivepath.com/publications/essays/archives/000361.php

O'Reilly, T. 2005. What is Web 2.0. Design patterns and business models for the next generation of software. O'Reilly, T, Sebastopol, CA. http://www.oreillynet.com/pub/a/oreilly/tim/news/2005/09/30/what-is-web-20.html

Robinson, J.B. 1990. Futures under glass: a recipe for people who hate to predict, *Futures* 22, no. 8: 820–842.
Roe Smith, M., and L. Marx, eds. 1994. *Does technology drive history?* Cambridge, MA: MIT Press.
Saul, P. 2006. Seeing the future in weak signals. *Journal of Futures Studies* 10, no. 3: 93–102.
Schnaars, S. 1989. *Megamistakes: forecasting and the myth off technological change.* New York: The Free Press.
Schwarz, B., U. Svedin, and B. Wittrock. 1982. *Methods in future studies.* Boulder, CO: Westview Press.
Shirky, C. 2005. Folksonomies & tags: the rise of user-developed classification. IMCExpo Conference, 14–23 April 2005.
Shuen, A. 2008. *Web 2.0: a strategy guide: business thinking and strategies behind successful Web 2.0 implementations.* Cambridge, MA: O'Reilly Media.
Surowiecki, J. 2004. *The wisdom of crowds.* New York: First Anchor Books.
Travers, S. 2007. Bubble 2.0: organized online critique of Web 2.0. Paper presented at Conference on 'Media in Transition 5: Creativity, Ownership, & Collaboration in the Digital Age', 27–29 April 2007, Massachussetts Institute of Technology, Cambridge.

Keeping the balance: exploring the link of futures research with innovation and strategy processes

Patrick A. van der Duin and Erik den Hartigh

Delft University of Technology, Faculty of Technology, Policy and Management, The Netherlands

Futures research can support decisions concerning innovation and strategy issues. To encourage companies to use the results of futures studies, the futures researcher has to be aware of how those companies (clients) have organised their innovation and strategy processes. It is our position that futures researchers should balance the way they conduct futures research with the way innovation and strategy processes are organised and carried out. For example, in an organisation with a hierarchical way of building a strategy, a 'democratic' and interactive approach to futures research will probably not be appreciated. This makes it interesting and worthwhile to find out what futures research and innovation and strategy have in common, and on which points they relate. This will help futures researchers improve their work and make sure that their work has an impact on decisions concerning innovation and strategy.

1. Introduction

Looking to the future has been given many different names, such as forecasting, technology assessment, futures studies, futures research and foresight. We use the term futures research and define it as 'the ability, the competence and the art of describing, explaining, predicting, exploring and interpreting future developments and its consequences, as the result of actions and decisions in the present' (Berkhout et al. 2007). This means that, in our view, futures research focuses on changes and developments rather than on singular events (e.g. Mendonça et al. 2004). Futures research can be carried out at different levels and from different perspectives. For instance, Pina e Cunha, Palma and Guimaraes da Costa view *foresight* from two different perspectives: (1) the focus is on the future or on the present, and (2) a microscopic or macroscopic analysis is carried out. Combining these two perspectives results in 'four different modes of organizational foresight: strategic planning, visioning, scenario thinking and planned emergence' (2006, 942). Based on our definition, futures research includes these four ways of looking to the future. Although in our definition we emphasise that futures research is also a kind of art, which may give futures research a somewhat intangible flavour, we would like to stress that we see futures research as an explicit and formal activity, which uses distinct methods, tools and processes, and that as a result we exclude more informal approaches to looking to the future.

Conducting futures research is not a goal in itself but serves a specific purpose, for example building a vision or testing a plan for the future. If the results of future studies are to be used, futures researchers would do well to understand how organisations have organised their strategy and innovation processes. For example, organising a bottom-up futures research process in an organisation that is centrally organised and managed is unlikely to have much of an impact. Similarly, a futures study that focuses on market-related aspects for a company that operates in a technology-driven industry will probably not be used. To *balance* futures research with decision-making concerning innovation and strategy, it is necessary to find out what connects futures research to strategy and innovation. That is the (explorative) goal of this article.

Indeed, there is a strong link between innovation and the future. Because innovation processes have to do with creating new products, services or processes, they provide ample opportunities to incorporate new (future) developments. A promising idea for an innovation that is based on an envisioned future can be supported or countered by future changes in technology, economy and/or society. Certain future expectations may turn out to be false and may have to be replaced by others. However, unforeseen future developments can make it more likely for an innovative idea to be realised. The implication of this is that innovators should take future changes into account and, subsequently, that every innovation process should take the future into account in an explicit way.

However, the numerous erroneous predictions of the market diffusion of innovations make it clear that the use of futures research in innovation processes is not without problems. A *technology push* approach, for example, ignores the influence of important societal and market-related developments on the acceptance of innovations. The risk involved in the *market pull* approach is that by only taking into account societal and market developments one limits the technological window of opportunity for innovation and runs the risk of missing out on new technological possibilities. It appears that it is essential for companies to balance carefully their innovation process and the way futures research is carried out and applied in this innovation process.

Strategy is closely related to the future as well. Because strategy essentially has to do with how a company sees its own future and how it will realise or be prepared for that future by taking different steps and decisions, every strategy process can benefit from futures research. Again, futures research should be in line with the strategy process in question. The diversity in terms of the types of strategy processes (e.g. outside-in, inside-out, static or dynamic) should be reflected in the corresponding futures research process. For example, the results of an interactive 'democratic' futures research process (i.e. one that involves many different stakeholders rather than a selected group of experts) may be hard to accept by strategists who are used to a top-down approach. Also, a study that is carried out by solitary experts or external consultants will have little impact on a strategy process involving many different stakeholders.

In Section 2 and its sub-sections, we provide a few definitions of strategy, and describe a specific perspective on strategy to connect it to the concept of the future. It will become clear that adopting different perspectives on strategy also means assigning different roles to 'the future'. In Section 3, we connect innovation (processes) to the future, showing that the future is approached differently in different phases of the innovation processes and in different innovation generations. We end this article with some concluding remarks not only about the link between innovation, strategy, and futures research but also on the relationship between the futures researcher and its client.

2. Strategy

Strategy means different things to different people. In this article we deal with corporate strategy, as distinct from, e.g. military strategy. Mintzberg (1994) mentions four basic definitions of (corporate) strategy: (1) strategy as a plan, (2) strategy as a pattern, (3) strategy as a position, and (4) strategy as a perspective. Every one of these definitions has a distinct relationship to the future. Corporate strategy as a *plan* is almost by definition related to the future of a company, as a plan is highly likely to contain actions or guidelines that are to be carried out to achieve future objectives. Corporate strategy as a *pattern*, i.e. a more or less consistent behaviour over time, may be related to the past as well as the future. It is possible to study the pattern of a company's strategy in the past, indeed one may well argued that studying the past is the only way to identify the pattern. A pattern, however, is also something that continually emerges from the past, present and future actions of the company. Further, a company's strategic pattern is shaped in the future. Corporate strategy as a *position* seems to be mostly related to the present. Nevertheless, we can study how past actions have resulted in the current position and we can analyse or plan how future actions may lead to a future position. Corporate strategy as a *perspective* is related to the present and the future. In the present, a perspective is related to the company's mission statement or fundamental purpose. With regard to the future, the perspective is a vision of what a company wants to be.

Given the many relationships between corporate strategy and the future, actually providing a complete overview of how *strategy* relates to *future* would be a task of gargantuan proportions that is outside the scope of this article. What we propose, therefore, is to provide an illustration of how various strategic schools of thought deal with the future. We examine the difference between static and dynamic approaches and the role of the future, and discuss the tools that are used to analyse the future.

2.1. *Strategy perspectives*

Researchers and managers can adopt different basic perspectives with regard to their subject matter. These basic perspectives are known as *theories of the firm*, because they provide explanations about the nature of the firm, i.e. why it exists, how it behaves and what its relationship to the market is. In this article we look at a few of the most important *theories of the firm*. We divide them on the basis of two dimensions:

(1) Does the reasoning start from the market or from the company? In orther words, is the future influencing the company or is the company influencing the future?
(2) Is the reasoning basically static or dynamic? That is, is the future predictable in a linear way, or is the future created on a continuous basis?

In doing so, we use the perspectives as defined by Teece and Pisano (1994) and we adopt the classification proposed by Zegveld (2000), see Figure 1.

The first dimension we use to distinguish theories of the firm is on the basis of their reasoning: starting with the market, i.e. outside-in, or starting with the company, i.e. inside-out.

2.2. *Static, outside-in*

Starting with the market implies taking an outside-in approach. The best-known perspective of this approach is the *industrial organisation theory of the firm* (Bain 1959; Scherer and Ross

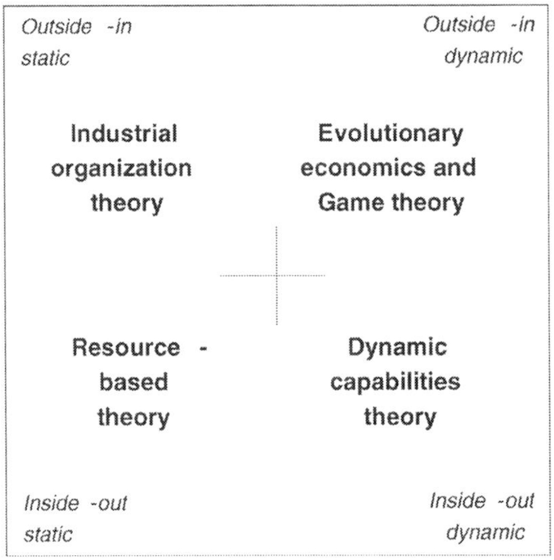

Figure 1. A classification of strategy perspectives (source: Zegveld 2004).

1990). The basic reasoning here is that the structure of a market determines the behaviour of the companies in that market, and that competition between the companies determines their performance. Companies make a deliberate choice in which market they want to compete and they try to establish a favourable position within their chosen market. Such a position will enable them to realise a cost advantage relative to their competitors, i.e. *cost leadership*, or to differentiate themselves from their competitors, i.e. *differentiation*.

To make strategic decisions using the industrial organisation perspective, it is important to have information on future market developments and on the way a company can compete given these market developments. Important tools in this respect are (see also Zegveld 2004):

- Industry analysis, i.e. how will the attractiveness of an industry develop in terms of market size, market growth, industry concentration and profit margins;
- Technology analysis, i.e. how will technology influence the structure of the and the company's cost functions;
- Futures analysis, i.e. analysis of market trends and patterns, environmental scanning and forecasting;
- Porter's 5-forces model, i.e. how will the bargaining position of the company change relative to that of suppliers, buyers, substitutions and new market entrants;
- SWOT analysis, i.e. what will the Opportunities and Threats in the market be and how can the company act upon these, given its own Strengths and Weaknesses;
- Value chain analysis, i.e. the activities that have to be carried out given the company's products and positioning.

When the future has been more or less accurately 'predicted' using these tools, management can decide what actions the company will take to become as profitable as possible, i.e. the industry segment it wants to be in, its strategic positioning, the products and services it will market and the investments it will make, e.g. the resources it needs to acquire, capabilities it needs to build. It is important to recognise that the reasoning of the industrial organisation perspective is essentially *static*: future market developments are predicted, a strategic plan is made, actions are taken and finally the outcomes become visible in the company's performance, i.e. profits, cash flow or turnover. In the meantime, it is assumed that the market does not change in discontinuous ways.

2.3. *Static, inside-out*

Starting with the company implies taking an inside-out approach. The best-known example involving this approach is the *resource-based theory of the firm* and its more modern successors like the *core capabilities* or *core competence* perspectives (Penrose 1959; Wernerfelt 1984; Prahalad and Hamel 1990; Barney 1991). In this theory the starting point for reasoning is a company's strategic intent or ambition. Based on this ambition, the company builds competencies and accumulates resources that are unique and difficult to copy by its competitors. With these competencies and resources it will do different things, like carrying out R&D, and introducing and selling products and services in the market. Through these actions, the company can fulfil existing or new customer needs, thereby capitalising on an existing market or creating a new one.

When making strategic decisions based on the resource-based perspective, it is important to have information concerning a company's strategic intent, its assets and capabilities and the way these can be deployed to the company's maximum advantage. Important tools in this respect are (see also Zegveld 2004):

- Strategic intent analysis, i.e. what are a company's long-term ambitions;
- Technology analysis, i.e. which technologies are available that can be favourably developed by the company, can different technologies be combined to create architectural innovations;
- Competency analysis, i.e. what are the company's most important competencies, how can they be deployed, which ones are missing given the ambitions;
- Asset analysis, i.e. what are the company's most important resources, e.g. technologies, brands, knowledge, cash, and what is their value, how can they be deployed and which ones are missing given the ambitions;
- Strategic architecture analysis, i.e. how can the company's assets and capabilities be combined and configured in a logical way to realise the ambitions that have been defined;
- Visioning, backcasting, environmental scanning.

When the company's strategic intent has been determined and the resources and capabilities identified, management can decide on which existing resources and capabilities to build, which new ones to acquire or invest in, and subsequently which activities it will carry out, which customers it will target and which markets it will serve. Here, too, we should realise that the approach is (again) essentially *static*: once the plans have been made, strategic intent is not supposed to change in major ways and the value of assets and capabilities is not supposed to change abruptly.

In both perspectives discussed above, outside-in and inside-out, it is *assumed* that there is one, relatively predictable future. Of course, no management scientist is naive enough to believe that the future is indeed one-dimensional and predictable, but in these perspectives it is assumed to be at least predictable enough for the company to implement its strategy, test its effects on performance and when necessary change it. It is highly questionable whether in today's turbulent world there is any market or situation where such predictability exists and yet, many corporate strategic planning processes still build on these static perspectives. Also many management scientists still dominantly use those perspectives for their research. An example is an article by Antoniou and

Ansoff (2004, 275) in this journal. While they acknowledge that '... high technology turbulence is going to increase ...', the whole article is still built on ideas of forecasting and then determining specific actions and priorities to determine the organisations future direction. We think that an important reason why management scientists still take these approaches is that designing and analysing company strategy becomes progressively more difficult when we take dynamics and uncertainties into account. We know that the use of a static model will at least provide us with some basic information and we are all too happy to put off the dynamic aspects of our corporate strategy to the implementation phase. Part of the problem is that frameworks and methods for dynamic planning or for dynamic research are not yet fully developed.

2.4. *Dynamic, outside-in*

In the dynamic perspectives, too, we draw a distinction between the inside-out and outside-in approaches. Evolutionary economics and (evolutionary) game theory are the main perspectives in which an outside-in approach is used and dynamics are taken into account explicitly (Hannan and Freeman 1977; Nelson and Winter 1982; Levinthal 1991; Brandenburger and Nalebuff 1995).

Scholars using evolutionary economics study the evolution of business populations and the ways in which individual companies manage to survive the environmental selection forces by adapting to their environment. In this perspective, the future is not a static thing to which companies can easily adapt, but it is fickle and changing. It requires not *adaptation*, but rather *adaptability*. Companies that are unable to adapt continually will be forced out of business.

Game theory, especially in its evolutionary variant, is a perspective in which scholars analyse the way interaction between individual players affect the collective outcomes of this interaction. This implies that future outcomes are not seen as uniquely predictable, but as depending on the actions of the players in the game. For example, in the well-known *prisoner's dilemma*, a single-round game of two players that can each have two choices; there are $2 \times 2 = 4$ possible futures. Increasing the number of rounds in the play, i.e. evolutionary game theory, increasing the numbers of players or increasing the numbers of actions each player can take, exponentially increases the number of possible futures. The corporate strategy interpretation of game theory starts from the *value net*, which provides an identification of the players and the interactions between them. The next step is to identify the elements of the game and find opportunities to change the game, e.g. by changing the players, the rules or the scope of the game.

To make strategic decisions on the basis of these perspectives it is important to have an idea as to how interaction between players shapes the future and of how this future in turn influences the possible actions of the company. Important tools in this respect are (see also Zegveld 2004):

- Population ecology analysis, i.e. analysing the forces of selection that the environment imposes;
- Adaptability analysis, i.e. how the company can adapt to its environment and how it can increase its adaptability;
- Value net analysis, i.e. analysing the players, identifying the main categories into which they fall, and how they are related;
- Cooperative game analysis, i.e. analysing the dynamics of the value net when players cooperate;
- Non-cooperative game analysis, i.e. analysing the dynamics of value net when players compete;
- Scenario analysis.

These tools and the research fields from which they stem have become increasingly popular. In organisation population ecology and selection, research has been conducted into the way the environmental selection processes work. A particularly promising approach is that of computational analysis and agent-based computer simulation (Epstein and Axtell 1996). These relatively novel methods provide relatively easy ways to explore multiple futures and to examine ways in which a company can adapt to and even influence or create these futures. The most important challenge in this field is to create a deep managerial understanding to match the analytic rigour of the methods involved.

From the field of organisational learning, another promising stream of analysis has emerged, namely that of scenario planning (De Geus 1997; Van der Heijden 1996). Experience with strategic planning at Royal Dutch/Shell made it clear that conventional strategic planning methods like those advocated by Ansoff were no longer sufficient. In the 1980s, Arie de Geus and Kees van der Heijden at Royal Dutch/Shell's strategic planning department started experimenting with scenario planning, no longer assuming a single future for the company, but rather a portfolio of different futures.

2.5. *Dynamic, inside-out*

Since the beginning of the 1980's, considerable research has been carried out using the resource-based view. While the resource-based view was still in its infancy (e.g. Newbert 2007), a need for a more dynamic version appeared. The article *Competing for the future* by Hamel and Prahalad (1994) already pointed in this direction. In 1994, Teece and Pisano coined the term *dynamic capabilities*, in which they emphasise two 'key aspects': (1) the shifting character of the environment, and (2) '... the key role of strategic management in appropriately adapting, integrating, and re-configuring internal and external organizational skills, resources, and functional competences toward changing environment' (1994, 538). The dynamic capabilities perspective is an extension of the resource-based view, which, while still departing from the company's capabilities, explicitly takes into account the paths that have created these capabilities and the future paths that are open to the company (Teece and Pisano 1994; Eisenhardt and Martin 2000; Helfat and Peteraf 2003). Core notions are dynamics, evolution, change and path-dependence. Companies should develop dynamic capabilities, which means that they have to be able to renew existing resources or competences. In this way, they will become adaptable and will therefore be able to address a continuously changing environment. This is difficult because of *path-dependence*: current capabilities and positions are heavily dependent on existing resource commitments. Similarly, commitments made today will limit the possibilities that are available at a future time.

In this perspective, the future is seen as being shaped in the interaction between a company and its environment. As far as the concept of future is concerned this means that, on the one hand, there is no single, predictable future, while on the other hand, path-dependency ensures that the range of possible futures is not endless. To make strategic decisions using the dynamic capabilities perspective, it is important to have information regarding a company's existing resources and competences and how these can be changed or expanded, on the company's positions, on how these positions came into being and on how they can be extended, and on the paths that are open for the future. Important tools in this respect are (see also Zegveld 2004):

- Strategic intent analysis, i.e. what are the company's long-term ambitions?;
- Dynamic capability analysis, i.e. analysing the company's current capabilities, how they can be deployed and adapted or changed when the environment so requires;

- Path-dependency analysis, i.e. analysing how resource commitments and investments today will influence the range of possibilities for the company's future;
- Technology regime analysis, i.e. which available technologies can be favourably developed by the company and what are the expected technology trajectories?;
- Company innovation system analysis, i.e. how are the company's capabilities related to each other and to capabilities outside the company, or, when one capability is adapted, how will this require other internal and external capabilities to change?;
- Visioning, backcasting, environmental scanning.

With regard to some of these tools and methods, such as adaptive capabilities, path-dependence and technology regime analysis, an increasing body of knowledge is becoming available. Such research provides the basic knowledge needed to understand how, on the one hand, the future can be understood as dynamic and inherently unpredictable, and how, on the other hand, it can be understood as limited in possibilities, because markets and companies can become locked in to a single technological regime or be stuck with their existing capabilities for extended periods of time.

To summarise the place of 'the future' in strategic thinking, Figure 2 illustrates which place the future has in different strategic perspectives. A static perspective is in line with a stable and (thereby) predictable future, while outside-in is associated with more change and with much less predictability. Static and dynamic refers to the source of change and thereby uncertainty which can be located within the organisation or within its environment.

We have seen that the static perspectives on corporate strategy are becoming less and less relevant in today's dynamic world. This is true for academic analysis as well as for practical strategy making. The dynamic perspectives on corporate strategy have, until now, received considerably less attention than the static perspectives. The body of knowledge as relevant to scholars or managers is indeed growing, but is still relatively small. This causes many scholars and managers to stick to the well-known static perspectives. There is therefore a huge need to adopt perspectives that

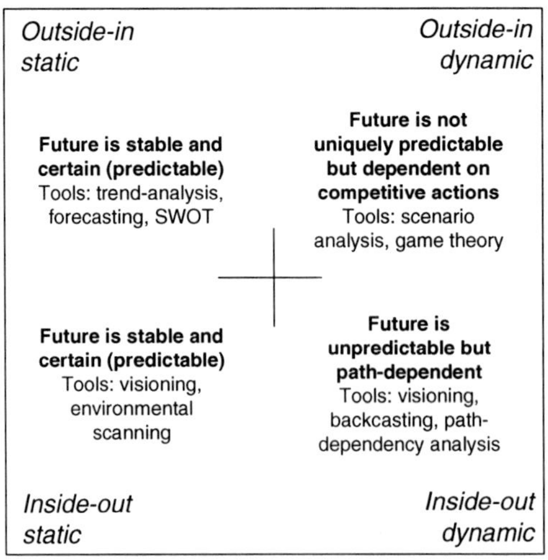

Figure 2. The place of the future in strategic thinking.

recognise not just one, predictable future, but a whole range of possible futures. This also means the application of more advanced tools. This need for more dynamic approaches of corporate strategy was identified by scholars from Wharton School in 1997 (Day, Reibstein, and Gunther 1997), but it still holds today.

3. Innovation

Many studies show that innovative companies have higher profits and profit margins than companies that innovate to a lesser degree or not at all (e.g. Jonash and Somerlatte 1999). Nevertheless, innovation has a long history and it can be considered one of mankind's oldest activities. From a modern perspective we see that the first R&D-laboratories were established in nineteenth-century Germany. These laboratories were part of the chemical industry and they were soon followed by similar laboratories in the USA (Bassala 2001). The current wave of attention to innovation from management scientists, companies and governments began shortly after the Second World War, when Vannevar Bush, the then Director of the Office of Scientific Research and Development, presented his influential report entitled *Science, The Endless Frontier* (Bush 1945). In this report, Bush states that scientific research is the basis of technological growth, and therefore of economic growth and an increase in military power. Bush's ideas were the start of a series of different innovation management *practices*, each of them dominant in a specific historical period. These four *generations* of innovation management (based on Liyanage, Greenfield, and Don 1999; Miller 2001; Roussel, Saad, and Erickson 1991) are shown in Table 1.

This historical development of innovation management can be characterised as *evolutionary*, with every new generation being adjusted to its economic and societal environment in an attempt to remedy the drawbacks of the generation that came before, making innovation more complex (Ortt and Van der Duin 2008). A future perspective on this development via different generations may be motivated by a need to look for clues for a next generation. Although some authors claim that there is a fifth generation (Rothwell 1994), their attempts to innovate innovation management are insufficient to speak of a new innovation management *paradigm*. Moreover, many large companies have multiple and different types of innovation processes within their own organisation (Verloop 2006; Van den Elst, Tol, and Smits 2006). That is, companies may decide to apply (existing) innovation management principles on the basis of their organisational needs, thereby creating a kind of *contextual innovation management* (Ortt and Van der Duin 2008; see also Gebhardt 2005).

Table 1. The four generations of innovation management.

Generations	Principles of innovation management
1950 – mid 1960s	Technology push, linear innovation processes
Mid 1960s – early 1970s	Market pull, linear innovation processes
Early 1970s – mid 1980s	Combined market pull and technology push, coupled innovation processes (with feedback loops)
Mid 1980s – early 2000s	Innovation in networks and systems, 'open innovation' (Chesbrough 2003)

Innovation involves coping with future uncertainty because an innovation will be marketed at a future date when new developments may have changed the market situation. One should be aware that innovation processes take time, more often than not a great deal of time. In spite of the rhetoric surrounding shorter *time-to-market*, the overall development of an innovation still takes a lot of time, and often the time needed is not decreasing. The development time varies, of course, with each type of innovation. An incremental innovation, i.e. a slight change to or improvement of a product or service, such as a new version of specific software, may take up to a couple of months or a year. However, a more radical innovation that involves changes in many aspects of the innovation, can take much longer. It takes 7 years, for instance, to develop a new car, and between 10 and 15 years in the case of a new drug. As a result of this, a *potentially* good idea for an innovation may turn out to be a bad idea when the innovation is actually marketed. That is to say, between the first idea for an innovation idea and its future implementation, society, market, technology and economy may have changed in such a way as to render the originally envisioned benefits obsolete (Van der Duin 2006; see also Pohlmann, Gebhardt, and Etkowicz 2005). Future changes may also support the original innovation idea. In either case, innovation and future are closely connected, almost in a natural way.

Innovating, i.e. developing new products and services, involves breaking with old habits and thereby getting rid of a company's old portfolio. In a way, the future, or at least a company's future, is being shaped by all those innovations. Taken together, innovations form a kind of *transition path* to an envisioned future. This means that one future is predicted and used as a kind of *bait*. The image of the future is meant to inspire innovators (e.g. Könnöla, Brummer, and Salo 2007), and to serve as a kind of goal for which they should aim. Innovations are then designed to answer the question as to how the envisaged future must be realised. The question can also be put differently: is the innovation we are developing suitable for different futures? This question is asked from a more modest stance with regard to the future. People who ask this question have less faith in shaping the future by themselves or their organisation, but instead see the future as a given, not by themselves but by other stakeholders.

The close link between futures research and innovation or, more specifically, the balance between innovation and future research, is clearly illustrated by the historical development of these two topics, as is shown in Table 2 (from Van der Duin 2006).

Table 2 shows that, at a general level, innovation management and futures research have developed in parallel. It is difficult to determine which discipline influenced the other, but the resemblance is so striking that it can be assumed that futures research can be used more effectively in innovation when the two disciplines are mirrored. In other words, an innovation process should use a technology forecast, and in an innovation system futures research should be an 'open' process involving many different kinds of input from futures researchers of the various

Table 2. Generations of innovation management and futures research.

	Innovation processes	Futures research
Generation 1: 1950–1960	Technology push	Technology forecasting
Generation 2: 1960–1970	Market pull	Technology assessment
Generation 3: 1970–1980	Coupled innovation processes	Explorative futures research
Generation 4: 1980–present	Innovation in systems or networks	Networked or systemic futures research

organisations involved. As a consequence, the concept of *contextual innovation* can also be applied to futures research: *contextual futures research*, whereby the decision as to how to carry out futures research is determined by the type of innovation process being used.

Innovation management involves the general principles of addressing innovation at an organisational level. At a lower level of detail there are innovation *processes*, where the specific activities of an innovation *project* are described. Often, these innovation processes are pictured as a linear process consisting of a number subsequent phases. In each phase innovation futures research (or technological forecasting) plays a different role. Table 3 is from Twiss (1992) and illustrates the link between futures research (i.e. (technology) forecasting) and innovation at a process level.

In Table 3 futures research is predominantly represented as forecasting. In the first two phases, i.e. idea generation and technical feasibility, forecasting plays its most important role. These phases are the farthest removed from market implementation, which makes looking at the future valuable. Because little (quantitative) information is available in these phases, predictions are not very accurate. Fortunately, getting it wrong at this stage will not be very costly. Given that the level of investments in the innovation increases when its development process comes closer to market implementation, accuracy will be increasingly important. Phases 3, 4, and 5 benefit less from forecasting because at that point most decisions about the innovation have already been made, which means that these phases are more operational than strategic in nature. To summarise, futures research in general and forecasting in particular will be more important in the early phases of a linear innovation process than during the later stages, when market research methods will be more popular.

Lichtenthaler (2007) views the link between technology intelligence processes (TI) (which we consider a special kind of futures research) and (specifically) radical innovations from a more organisational perspective (or level). He distinguishes three different types of technology intelligence processes: (1) the hierarchical technology intelligence process, (2) the participatory intelligence process, and (3) the hybrid technology intelligence process. Each of these three types influences R&D and innovation processes. For instance, in type (1) future trends that may be relevant to technological research and innovation directions are directly communicated to top management; or, in type (2), although trends are identified at an early stage of the innovation, they are (nevertheless) debated quite heavily among different management levels and can even lead to political conflicts; and in type (3) decisions concerning R&D and innovation are 'mainly based

Table 3. The relationship between forecasting and the technological innovation process (Twiss 1992).

	Technology forecasts		
Phase of the innovation process	Importance	Accuracy	Financial effect of forecasting error
1. Idea generation	High	Medium	Low
2. Technical feasibility	High	Medium	Low
3. Design and development	Low	High	Medium
4. Preparation for production and marketing	Very low	High	High
5. Post launch	Not applicable	Not applicable	Not applicable

on the assessment of the researcher and partly based on analyses of the TI specialist or external experts' (Lichtenthaler 2007, 112).

Of course, the organisational level and the process level of innovation and, subsequently, the way futures research is carried out, do not exist in isolation. This means that the innovation process outlined by Twiss (1992) is a first generation type of innovation process and will probably be more popular in Lichtenthaler's type 1 organisations than in the other two types. Managers of type 1 organisations will probably not like a second generation type of innovation process, in which market factors are considered very important and in which future trends will tend to be market-related or societal.

4. Discussion

The use and impact of futures research is decided not only on the basis of its quality, i.e. the way it is carried out. It is equally important to connect futures research to how a company is strategising and innovating. The worst thing a futures researcher may hear is not that his or her work is not up to scratch but that it does not relate, either in content or process, to the company involved and the decisions it is about to make. This means that improving futures research not only involves improving the methods and processes involved, but that it is equally important to connect futures research to the decision-making processes of the companies involved. However, futures researchers need to understand not only the *immediate* context of their work, i.e. the decision-making process as such, but the *organisational* context as well. A better conceptual understanding of the principles of strategy and innovation can be seen as a fruitful way to improve the quality of futures research.

In the discussion on the role of the future in strategy and innovation, we saw that there is a great need for a further development of dynamic perspectives, not just as fancy new tools for academic analysis, but also as practical tools for managers. Futures research that on the one hand is analytically rigorous and that on the other hand interacts with strategy directly seems extremely suited to connect to such developments. In this way, futures research may offer a valuable contribution to dynamic views on strategy, in which the future is seen as changeable and multi-dimensional rather than static and predictable.

In the discussion on the role of futures research in innovation, we saw that innovation and the future are related in an almost natural way, whereby the future can serve as a source of inspiration to innovators. The close relationship between futures research and innovation is also illustrated by their almost parallel historical development. At an innovation process level, futures research adds the most value in the initial phases, when little information about the innovation is available and when the innovation is still a long way from being implemented. A focus on the future and a focus on innovation can, therefore, be seen as two sides of the same coin.

To see how our opinion about aligning futures research with strategy and innovation management matches with reality, we performed an analysis of 45 multinational companies in the field of IT, media and telecom. The analysis was performed by letting 45 trained master students of a management programme each analyse a company according to different management models. They were specifically instructed to perform strictly fact-based analyses, i.e. not giving their own beliefs or opinions. We asked them to score for 'their' company whether over the past 5 years the company's internal environment (the value chain) had significantly changed and whether over the past 5 years the company's direct external environment (the value system) had significantly changed. The results are displayed in Figure 3.

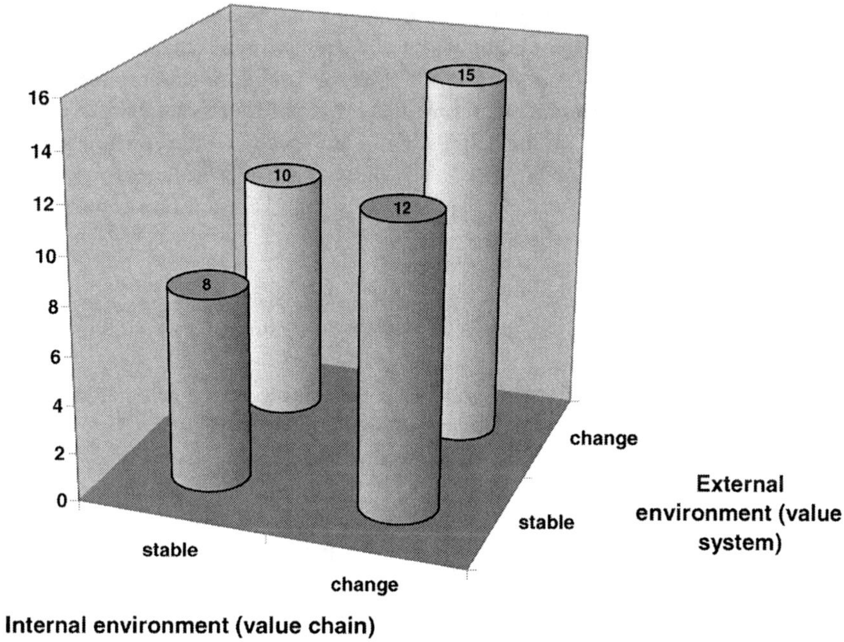

Figure 3. Stability or change in companies' internal and external environments.

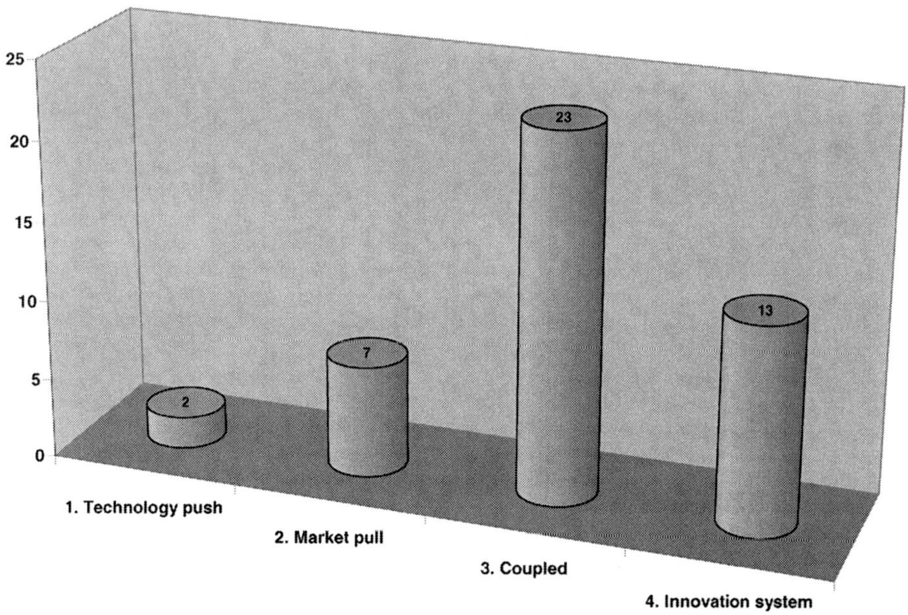

Figure 4. Companies' innovation generations.

From these numbers we can conclude that the internal and the external environment could be classified as relatively stable only for eight of the companies. In all other cases there is strong dynamism in either the internal or in the external environment or in both. In one-third of the cases both the internal and the external environment show significant dynamism. In our view, this means that there is a strong need for the application of advanced methods of futures research in these companies' strategy processes. We specifically think of methods fitting the dynamic strategy view, e.g. scenario analysis, (qualitative) game theory, visioning, backcasting or path-dependency analysis. Hence we formulate hypothesis 1:

H1: The more dynamic the internal and external environment of the company, the more dynamic its view on strategy, and the more it will use advanced methods of futures research in its strategy process.

Regarding innovation, we analysed for the same 45 companies to which generation of innovation management they belong. Again we asked trained master students to perform a fact-based analysis of the company's innovation processes, and draw a conclusion based on that analysis. The results are displayed in Figure 4.

From these numbers we conclude that the majority of companies apply innovation processes of the third or fourth generation. In our view, this means that there is a strong need for the application of advanced methods of futures research in these companies' innovation processes. We specifically think of the methods fitting the third and fourth innovation generation, i.e. explorative futures research or networked/systemic futures research. Hence we formulate hypothesis 2:

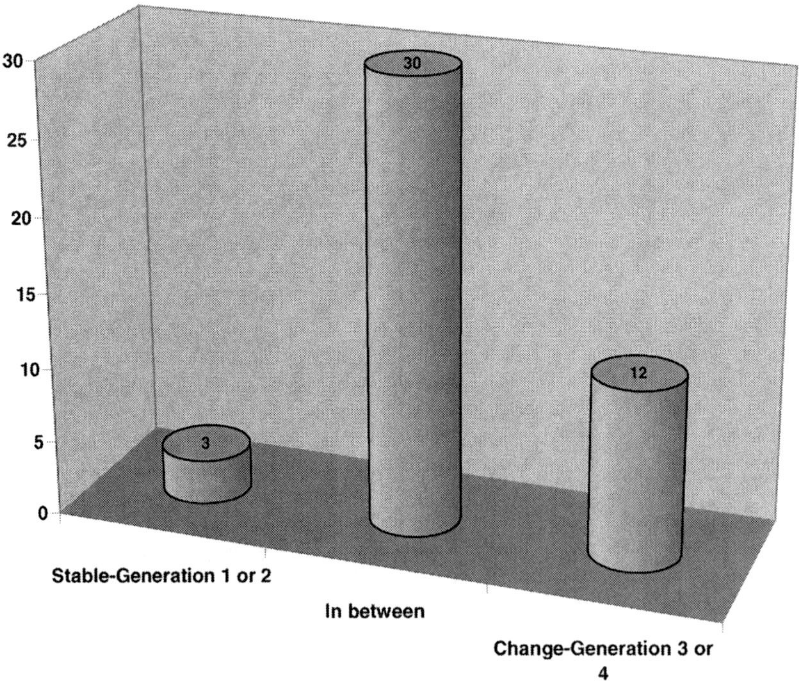

Figure 5. Companies' stability, dynamism and innovation generation.

H2: The more advanced the company's innovation generation, the more it will use advanced methods of futures research in its innovation processes.

When we combine the findings for the company's strategic stability and the company's innovation generation, we get a picture that points even more strongly towards the need for advanced futures research methods. We analysed how many of the 45 companies have a stable internal and external environment and belong to innovation generation 1 or 2. These companies can be qualified as strategically stable and innovatively less advanced. Such a situation applies for only three out of the 45 companies. Then we analysed how many of the 45 companies have a dynamic internal and external environment and belong to innovation generations 3 or 4. These companies can be qualified as strategically dynamic and innovatively advanced. Such a situation applies to 12 out of the 45 companies. All other companies (30) have either a dynamic internal or external environment and/or an advanced innovation process. These results are displayed in Figure 5.

Below we give two examples (Boxes 1 and 2): one of a company with a relatively stable environment that applies an earlier-generation innovation model, namely Canon, and one of a company with a very dynamic environment that applies a later-generation innovation model, namely Nortel Networks.

Looking at Canon and Nortel, we clearly see that the company with the more dynamic environment and the more advanced innovation generation applies a more advanced model of dealing

Canon is active in many markets, such as copying machines, printers, cameras and lithography equipment. In our analysis we established that in the past 5 years Canon experienced a relatively modest dynamism of its internal and external environment. The activities in it value chain have not changed a great deal and the same is true for the composition of its value system. Change was mainly gradual, in line with the company's 'Kaizen' principles, rather than radical. Of course, the technologies in Canon's market are changing, but these changes do not require completely new configurations of the company value chain or value system.

Canon states that 'New technologies are the key to our success ...', that it '... intends to aggressively pursue the development of highly advanced technologies that can be applied in new business domains'. and that 'Canon is pursuing the development of innovative technologies that will support its push into new business domains'. Hence we classify Canon as a predominantly technology-driven company, with innovation processes that belong mainly to the first generation (technology push).

Given this positioning, we would expect Canon to use relatively static strategic plans to realise its strategic goals and to plan its innovation processes and, indeed, this is what we see. Canon employs a long-term overall strategic planning model to become a 'truly excellent global corporation'. In phase I (1996–2000) Canon focused on building a strong financial structure; in phase II (2001–2005) on becoming the number 1 in the world in all their major areas of business; and in phase III (2006–2010) on pursuing sustainability, innovation and growth. In other words, first make the basis, then reach the goals and then stabilise what was achieved. This kind of planning and forecasting process fits perfectly with an environment which is relatively predictable and with the technology-push orientation of the company.

Box 1. Canon.

Nortel Networks is offering products, services and solutions based on telecommunications and networking technologies for a variety of industrial and government clients all over the world. Since the beginning of the millennium Nortel changed its focus from (vertically integrated) manufacturing to exploiting proprietary technology and providing solutions. This means that also the activities in its value chain changed dramatically, e.g. manufacturing activities are drastically outsourced. Hence the internal environment can be qualified as dynamic. It also meant forging new relations with manufacturing partners. Add to this the dynamism of the market environment in terms of technology convergence and changing customer demands (e.g. more individual solutions), and it is clear that also the company's outside environment can be qualified as dynamic.

Nortel's innovation processes are built on the combination of the principles of technology leadership and market leadership. It combines R&D centres of excellence developing the technology with Nortel regional groups providing the market/customer perspective, and integrates these perspectives in its operations, where tailored solutions and services for customers are developed. This seems to qualify as a typical third generation innovation process, the coupled innovation model.

Given this positioning in strategy and innovation, we would expect Nortel to apply dynamic strategic thinking and advanced methods of futures research. After years of very weak financial performance, the new CEO, Mike Zafirovski, provided the company with a new philosophy 'Business made simple'. This is explicitly put forward as a philosophy, not as a fixed strategic plan. While we cannot look into the company to see what kind of futures research they apply, we can detect clues that they are applying dynamic and advanced thinking, e.g. they acknowledge that business comes from interaction between company and customers, they develop and provide customer-specific solutions, they demand increasing flexibility from the workforce, they apply decentralised, less hierarchical models for decision-making and innovation and they state a flexible, dynamic strategic vision. This kind of future thinking fits with a dynamic internal and external environment and with a third-generation innovation model.

Box 2. Nortel Networks.

with the future. Of course, these are just examples. Further research would be necessary to see whether our hypothesis could be confirmed.

5. Concluding remarks

We end this article by stressing that balancing futures research with innovation and strategy does not mean that decision-makers, strategists or principals of futures research-projects decide what the outcome of these projects should be. It is necessary to connect futures research to innovation and strategy (mainly) at a *process* level, and not at the level of content. It is important for futures researchers to maintain their independence and not yield a kind of 'official future' that simply reflects the dominant way of thinking within a company, often the personal opinion of a high-ranked manager. In this way, futures researchers avoid becoming puppets that simply support decisions that have already been made (informally).

Although futures researchers should avoid presenting a future that has no relationship with a company's current position, it may be even more dangerous to support decisions that have already been made. Given the fact that the future is different from the present (and the past), studying the future should always yield new information, new insights, and even new knowledge about the future. Simply confirming decisions that have already been made may only please the principal or main decision-maker, but a majority of an organisation's employees will not accept this, which may seriously undermine the futures researcher's credibility.

The futures researcher should keep this in mind not only during the project but especially before the project starts, since in that phase the futures researcher can ask its potential client if he or she is willing to accept an outcome that might not be what the client would like to hear. We do not mean doomsday scenarios or stories about the end of the world but studies of the future that simply do not support the decision a client may prefer. The futures researcher should only accept an assignment if he or she is convinced that the client will even accept an outcome that is not what he or she expected or would like to hear. This may mean that the assignment is not given, which will certainly have financial consequences, but the futures researcher should consider this a short-term problem.

If futures researchers adopt a moral position in this respect, they will maintain their credibility, which in the long run will prove to be a better strategy than opting in favour of short-term gains at the expense of long-term success. After all, what company can really be interested in hiring a futures researcher who is only in it for the short-term financial benefits? Practice what you preach is also a valid maxim for futures researchers.

So, the 'battle for the future' is on and it can only be won if futures researchers can clearly demonstrate the added value of their methods by connecting what they do to their clients' organisation and at the same time maintaining their independence. This is the only way in which futures researchers can help generate sustainable competitive advantage, for their clients as well as for themselves.

References

Antoniou, P.H., and I.H. Ansoff. 2004. Strategic management of technology. *Technology Analysis and Strategic Management* 16, no. 2: 275–291.

Bain, J.S. 1959. *Industrial organization.* New York: John Wiley.

Barney, J.B. 1991. Firm resources and sustained competitive advantage. *Journal of Management* 17: 99–120.

Bassala, G. 2001. *The evolution of technology.* Cambridge: Cambridge University Press.

Berkhout A.J., P.A. van der Duin, L. Hartmann, and J.R. Ortt. 2007. *The cyclic nature of innovation: connecting hard sciences with soft values.* Oxford: Elsevier.

Brandenburger, A.M., and B.J. Nalebuff. 1995. The right game: use game theory to shape strategy. *Harvard Business Review* 73: 57–71.

Bush, V. 1945. *Science the endless frontier.* Washington, DC: United States Government Printing Office.

Chesbrough, H.W. 2003. *Open innovation. The new imperative for creating and profiting from new technology.* Boston, MA: Harvard Business School Press.

Day, G.S., D.J. Reibstein with R.E. Gunther, eds. 1997. *Wharton on dynamic competitive strategy*. New York: John Wiley.
De Geus, A. 1997. *The living company*. Boston, MA: Harvard Business School Press.
Eisenhardt, K.M., and J.A. Martin. 2000. Dynamic capabilities: what are they? *Strategic Management Journal* 21: 1105–1121.
Epstein, J., and R. Axtell. 1996. *Growing artificial societies: social science from the bottom up*. Cambridge, MA: MIT Press.
Gebhardt, C. 2005. The impact of managerial rationality on the organizational paradigm. Role models in the management of innovation. *Technology Analysis and Strategic Management* 17, no. 1: 21–34.
Hamel, G., and C.K. Prahalad. 1994. Competing for the future. *Harvard Business Review* 72: 122–128.
Hannan, M.T., and J. Freeman. 1977. The population ecology of organizations. *American Journal of Sociology* 82, no. 5: 929–964.
Helfat, C.E., and M.A. Peteraf. 2003. The dynamic resource-based view: capability lifecycles. *Strategic Management Journal* 24: 997–1010.
Jonash, R.S., and T. Sommerlatte. 1999. *The innovation premium. How next generation companies are achieving peak performance and profitability*. Reading, MA: Perseus Books.
Könnölä, T., V. Brummer, and A. Salo. 2007. Diversity in foresight: insights from the fostering of innovation ideas. *Technological Forecasting and Social Change* 74: 608–626.
Levinthal, D.A. 1991. Organizational adaptation and environmental selection – interrelated processes of change. *Organization Science* 3: 47–71.
Lichtenthaler, E. 2007. Managing technology intelligence proceses in situations of radical technological change. *Technological Forecasting and Social Change* 74: 1109–1136.
Liyanage, S., P.F. Greenfield, and R. Don. 1999. Towards a fourth generation R&D management model – research networks in knowledge management. *International Journal of Technology Management* 18, no. 3/4: 372–393.
Mendonca, S., M. Pinha e Cunha, J. Kaivo-oja, and F. Ruff. 2004. Wild cards, weak signals and organisational improvisation. *Futures* 36: 201–218.
Miller, W.L. 2001. Innovation for business growth. *Research Technology Management* 44: 26–41.
Mintzberg, H. 1994. The fall and rise of strategic planning. *Harvard Business Review* 72: 107–114.
Mintzberg, H. 2004. *The rise and fall of strategic planning*. New York: The Free Press.
Nelson, R.R., and S. Winter. 1982. *An evolutionary theory of economic change*. Cambridge: Harvard University Press.
Newbert, S.L. (2007). Empirical research on the resource-based view of the firm: an assessment and suggestions for further research. *Strategic Management Journal* 28: 121–146.
Ortt, J.R., and P.A. van der Duin. 2008. The evolution of innovation management towards a contextual approach. *European Journal of Innovation Management* 11, no. 4: 522–538.
Penrose, E.T. 1959. *The theory of the growth of the firm*. New York: Wiley.
Pina e Cunha, M., P. Palma, and N. Guimaraes da Costa. 2006. Fear of foresight: knowledge and ignorance in organizational foresight. *Futures* 38: 942–955.
Pohlmann, M., C. Gebhardt, and H. Etkowitz. 2005. The development of innovation systems and the art of innovation systems – strategy, control and the culture of innovation. *Technology Analysis and Strategic Management* 17, no. 1: 1–7.
Prahalad, C.K., and G. Hamel. 1990. The core competence of the corporation. *Harvard Business Review* May–June: 79–91.
Rothwell, R. 1994. Towards the fifth-generation innovation process. *International Marketing Review* 11, no. 1: 7–31.
Roussel, P.A., K.M. Saad, and T.J. Erickson. 1991. *Third generation R&D. Managing the link to corporate strategy*. Boston, MA: Arthur D. Little.
Scherer, F.N., and D. Ross. 1990. *Industrial market structure and economic performance*. Boston, MA: Houghton Mifflin.
Teece, D.J., and G. Pisano. 1994. Dynamic capabilities and strategic management. *Industrial and Corporate Change* 3: 537–556.
Twiss, B. 1992. *Forecasting for technologists and engineers. A practical guide for better decisions*. London: Peter Peregrinus.
Van den Elst, J., R. Tol, and R. Smits. 2006. Innovation in practice – Philips Applied Technologies. *International Journal of Technology Management* 34, no. 3–4: 217–231.
Van der Duin, P.A. 2006. *Qualitative futures research for innovation*. Delft: Eburon Academic.
Van der Heijden, K. 1996. *Scenarios: the art of strategic conversation*. Chichester, UK: Wiley.
Verloop, J. 2006. The Shell way to innovate. *International Journal of Technology Management* 34, no. 3–4: 243–259.

Wernerfelt, B. 1984. A resource-based view of the firm. *Strategic Management Journal* 5: 171–180.

Zegveld, M.A. 2000. *Competing with dual innovation strategies: a framework to analyse the balance between operational value creation and the development of resources*. Den Haag: Werk-veld.

Zegveld, M.A. 2004. Corporate strategy and the position of technology: a bird's eye view. In *Reader technology and strategy*, ed. M.A. Zegveld and E. Den Hartigh, 1–20. Delft: Delft University of Technology.

Changing perspectives on foresight and strategy: from foresight project management to the management of change in collective strategic elaboration processes

Sébastien Treyer

AgroParisTech/ENGREF, Management Sciences Department, Paris Cedex 15, France

Reflexivity about foresight activities for strategic management is often seen as too low. In order to enhance the reflexive capacity of foresight practice, a reframed perspective can be proposed, by considering individual foresight projects within the broader conceptual framework of 'future-oriented debates', analogous to the models of 'hybrid forum' or of 'strategic conversation' developed respectively in the fields of sociology of science and strategy, consisting both of the succession of foresight operations on a common theme, and of the community of actors associated to their discussion. Foresight can then be considered as an intervention in such a future-oriented debate, aiming at incremental change, with the objective to improve its quality, for which some criteria can be proposed. The usefulness of this change of perspectives from project management to the management of change is illustrated by the example of the international debate on future water scarcity.

Introduction

The practice of foresight plays an important role in science and technology policies (Martin 1995) as it is supposed to enlarge the scope of the science and technology strategy to the longer term future. Foresight is also linked more generally to the field of strategy, but it is very often practice oriented: it is seen as an operation able to produce a collective strategy for change at the scale of firms, other organisations, or a whole national innovation system. This paper argues that there is a lack of reflexivity in the field of foresight and that more useful links could be established with theoretical frameworks both in the fields of research in science and technology studies and in strategic management. It seems particularly important, as is made explicit below, to take into account that collective strategy, either in firms or for public policies, is the emergence of an ongoing collective strategic building process, rather than a specific product of an exercise producing the optimal strategy. If this perspective is accepted, foresight could be considered as an intervention aiming at changing and improving the conditions and structure of this collective strategy elaboration process. In such a perspective, foresight would relate much more to the management of incremental changes, than to mere project management of a single operation.

In the next section, this paper presents more precisely the reasons why a change of perspectives about foresight seems necessary and, in particular, where a certain lack of reflexivity about the practice of foresight projects might lie. The following section then defines and describes such a 'future-oriented debate', and makes some analogies with other conceptual frameworks existing either in management science or in the sociology of science. Foresight is defined as an intervention in such a 'future-oriented debate', which renews the approach for designing a foresight exercise in terms of intervention strategy. In the final section, the paper illustrates how such a conceptual framework makes it possible to set specific and pragmatic criteria for the evaluation of the foresight intervention.

In order to plead in favour of such a change in perspectives and to demonstrate the usefulness of the conceptual framework proposed, this paper takes an example from the field of water policies, thus extending the field of foresight for science and technology policies to include also the question of scientific expertise and the interactions between science (in this case, environmental sciences) and society, for collective innovation processes.

A change in perspectives about what foresight is, in order to be able to set more context specific objectives

When designing a foresight process, it is not easy to be clear and precise about the objectives that we try to reach. Of course, it is commonly acknowledged that foresight exercises have the following general aims:

- The official reason why many foresights are launched is that they should help decision makers take into account long-term evolutions in today's decisions, and help them design better strategies (Godet 1991; Martin 1995); this is mainly a cognitive objective.
- Many people, and even the same authors, also recognise that the former objective can in many cases be less important than the ability of foresight processes to wire up a community of persons involved in the exercise, so that they will be better prepared for collective action (Godet 1991; Reger 2001).

Once these very general objectives are stated, designing a foresight exercise can sometimes only consist of the implementation of standardised procedures. As an example, Michel Godet (1989, 1998) recommends for strategic foresights a succession of standard steps: systemic or cross impact matrix analysis, morphological analysis, scenarios of the evolution of the context, scenarios of possible strategies, analysis of the robustness of the strategies. Although the mere application of standard procedures is not a general feature of foresight activities, the temptation exists to search for more rigour in foresight by developing methods that can be reproduced in very different contexts and situations : improvements of a particular method should then be sought for by trying to make it more able to be generalised. In such a perspective, the quality of the foresight process is mainly measured either by the conformity of the procedure implemented to standards or to reproducible experiments given by literature, or by the fact that a generalisable procedure has been used. Not only might you like to evaluate a particular method and its ability to be reproduced, but also the specific performance of a foresight process in a given situation, which considerably changes the framework. Finally, in order to evaluate the foresight process for this purpose, one neeeds to be able to monitor:

- if the exercise really led to the design of a strategy and if this strategy was really proactive,
- and/or if the exercise enabled the building of a network of stakeholders that are committed to future collective action.

These long-term effects of foresights are very difficult to trace, however, which makes it often impossible to base methodological improvements on an analysis of such impacts.

This paper argues that the lack of a conceptual framework that can set more context-specific objectives is one of the main reasons why it is difficult to trace back the results, effects and impacts of foresight processes. It therefore proposes a conceptual framework in order to design, implement, and evaluate a foresight process with much more specificity to the context in which it should take place. It also proposes that such a framework will therefore lead to three important features:

(1) objectives of the foresight process can be more clearly and precisely set,
(2) the range of methodological possibilities can be reopened, in contrast with standardised procedures and tools,
(3) evaluation can be more pragmatic.

For this purpose a change in perspectives is proposed, from the most common point of view, mainly marked by a project management approach, where the foresight exercise is seen as the operation of creating or instituting from zero a reflection about the long-term future among a group of people, to a less common perspective where the foresight intervention is only aiming at an incremental change in an already existing debate about the future.

'Future-oriented debates'

Let us take an example from a specific field, namely, water resources management. When we look at what characteristics foresights have had in this field since the late 1960s, we mainly see that this unveils a diversity of possible foresight exercises, ranging from the mere extrapolation by an expert of a demographic curve illustrating the pace of reduction in the future of the amount of water resources available per capita at a country level (Falkenmark and Lindh 1974), to an integrated assessment exercise at the global level (*World water vision scenarios*, Rijsberman 2000) or to more local participatory exercises trying to build a common vision for the future with stakeholders and with the help of a water basin model (see for example Sreenath and Mesarovic 1997). For each of these foresight types, the general objectives stated in the former section may of course apply, but we mainly see that evaluating the impacts and the efficiency of each exercise can only be made with reference to its specific objectives.

It is also useful to consider that these very different exercises are all related to one another by the simple fact that they are all a part of a 'future-oriented debate' in the field of water resources management.

What this paper proposes to name at least temporarily a 'future-oriented debate' refers to the co-evolution process of:

- a corpus of representations of the futures in a specific field (be they curves, scenarios, modelling exercises, images, visions, etc.), each one constructed and elaborated with reference to the others;
- a community of persons and institutions associated to the elaboration and discussion of these representations of the future.

The use of the words 'corpus' and 'community' are intended to underline that we do not merely talk about a collection or juxtaposition of individual elements, neither in the cognitive dimension nor from the point of view of the persons involved, but that there are specific interactions among these individual representations of the future and the individual persons taking part in the discussion.

The example of foresight in water resources management will help us make these interactions more explicit. For a more in-depth analysis, the future-oriented debate on water resources management is presented in more detail in Treyer (2006) and, focussing particularly on the international debate, in Treyer (2005).

First, let us consider that the representations of the future of water resources management have been evolving from the 1960s until the present (2009) following the overall shape of the construction of a real corpus of references, in an analogy with the scientific corpus of an academic field of research, each new exercise building on the preceding state of the corpus.

Schematically, the first period of this 'future-oriented debate' mainly mobilised curves of evolution in the future of the water stress index (amount of renewable water resources available every year for each inhabitant of a country). As the demographic curve grows, the amount of water per capita diminishes, because natural renewable water resources are limited. Hydrologists use such curves in debate in order to enlighten us that future water crises are likely to take place and that making water available for every user in the future will become increasingly more diffficult. Such rough estimates of water scarcity in the future were at the same time efficient in raising the attention for water resources management but they were also seen as too simple.

It is therefore in reaction to these future water stress curves that some national water planners or water resources management specialists proposed other representations of the future of water resources management. The second period of the debate corresponds to the mobilisation of a more precise representation of the water resources and supply system. Specific projections are being made to represent the future of water supply (will new water resources be made available in the future and what do national water plans indicate?) and also the future of water demand (what is the foreseeable future growth of water demands for every water use, corresponding to demographic growth, to irrigation development plans, to the growth of industrial activities, and to the efforts at reduction of water losses in these various sectors?). Instead of the water stress index of the first period, the second period of the debate proposes rather to use a water exploitation index (see Margat 1992; Alcamo and Henrichs 2000), corresponding to the ratio between foreseen total water demand and the total water resources made available, which is supposed to better represent possible situations of water scarcity, when the index gets ever closer to 100%.

Although these new curves of the water exploitation index might better represent a national water scarcity situation, some other researchers from the field of economics, history or political science proposed a new representation of future transitions of countries towards ever scarcer water resources. They mobilised specific 'hydropolitics' case studies telling the story of countries that have been experiencing water scarcity transitions to insist on the necessity to complement the former mainly technical representation of the water resources and supply system with political and economic dimensions. For these researchers, the future transition towards water scarcity in semi-arid and arid countries cannot be interpreted correctly if one does not take into account the importance of the power relationships between the state and the water users (Ohlsson and Turton 1999). Based on past examples of some water scarce countries, they insist on the necessity to envision future water scarcity transitions as the process of transformation of the role of a national state from the responsability of being the water supplier to the role of regulator of the share of limited resources among competing water users.

Of course these three types of representations still coexist today in the debate about water resources management, but it is striking to see that the present understanding of the most relevant future issues concerning water management is the result of a process of discussion among different representations of the future, the new ones being built on and in reaction to the former ones. This process is what we propose to consider as the construction of a corpus of representations of the future.

The forum, in which the discussions of these representations of future water scarcity take place, is mainly made up of UN institutions and UN experts (UNESCO, UNEP, FAO, etc.) and hydrologists within the academic community. It is also very important to see that the community of people participating in the discussion is growing in relation to the diversification of the representations of the future that have been mobilised. The very rough representation of water scarcity used in the first period raised the attention of national water planners, who decided to intervene and propose an alternative framework for water accouting. In this second period the stress put on technical balances and on official national plans envisioning the most probable future made the intervention of political scientists, specialising in the study of national planning administrations, unavoidable and their insistence to take into account the political dimension of future transitions.

The example of future water scarcity is a good illustration of the existence of a future-oriented debate on a specific theme lasting over more than three decades and evolving over time through the construction and discussion of new representations of what the future might look like and the involvement of new participants in the debate.

Based on this example, it is proposed to look at a future-oriented debate as being made out of the succession of foresight processes and foresight products on a common theme. In this perspective, a foresight exercise can only be seen as a moment in this longer lasting debate about the future.

The two dimensions of the future-oriented debate (corpus of representations of the future and forum of discussion), as well as the processes of evolution of the debate that link these two dimensions, make this conceptual framework very analogous to the concept of a 'hybrid forum' proposed in the field of the sociology of science (Callon, Lascoumes, and Barthe 2001) to describe how the reframing of the representation of reality by science evolves over time in close interaction with the access of new groups of participants to the discussion of these representations. This analogy will play an important role in our ability to identify possible evaluation frameworks for foresights.

It is also important to make clear that, for formalised organisations like firms, the future-oriented debate is very close to what the field of strategy calls a 'strategic conversation', consisting of the succession of representations of the future being discussed more or less formally in the organisation, and of the various moments of discussion taking place about these representations of the future (Van der Heijden 1996). It is, for example, the perspective on the management of strategic change used by Clark et al. (2007), introducing various types of narratives or representations of the future that might be used in such a strategic conversation. It is also the perspective used by Berkhout (2006). He refers to a process of negotiation and interaction among individual visions or expectations that may lead to collective visions, and states about this process that 'visions of the future are both ubiquitous and context specific'.

For this field of research, it is important to consider that strategy emerges from a collective strategy elaboration process and cannot be built directly through an intentional procedure.

Foresight as an intervention in an already existing future-oriented debate

Building on the definition of future-oriented debates, a change in perspectives is now possible, assuming that when a foresight exercise has to be designed, there is generally already a future-oriented debate or a strategic conversation going on, and that the foresight exercise is generally not needed in order to create from zero a new debate, but to change the conditions of the already existing debate.

In such a perspective, a foresight project cannot be designed and managed as if nothing existed before. On the contrary, a foresight exercise can be considered as only a momentary part of a longer lasting debate, already existing before, and going to last long after the foresight exercise is finished. Designing and implementing a foresight project therefore should be seen as a clinical intervention in a future-oriented debate in order to improve the way it functions.

For this purpose, the main feature of the intervention does not consist of the implementation of already existing, standardised procedures and tools for foresight. The crucial moment of the foresight intervention lies in the strategic diagnosis of the state of the existing future-oriented debate at the time the intervention begins.

This strategic diagnosis consists of answering the two following series of questions:

(1) What are the past and current representations of future evolutions of the theme considered that are being mobilised in the debate? Should they be improved and how?
(2) What is the state of the forum? Who participates? What are the rules of access to the debate? What are the rules set for the discussion within the forum? Should they be improved and how?

It might well be that in some situations, the strategic diagnosis leads to the conclusion that there is no existing debate, which would mean that it might be necessary to launch this new debate. We propose considering this situation as only one extreme case of the state of the future-oriented debate on which the foresight specialist is intervening.

Based on this strategic diagnosis of the initial state of the future-oriented debate, one can set very context-specific objectives to the foresight intervention that has to be designed. These objectives can either be related to an improvement of the representations of the future mobilised in the debate, or related to an improvement of the outreach of the community of people participating to the debate or to an improvement of the rules of discussion or of access of the debate, or to an improvement in both dimensions.

To take again the example of the future-oriented debate on water resources management, after the second phase of the debate, some politicians made the diagnosis that the political dimension was lacking in the representations of the future mobilised in the existing debate at that time. However, within the same debate but at another period of time, the International Hydrological Programme of UNESCO, the World Water Council and some other institutions made the diagnosis in 1998, before the World Water Forum of The Hague, that the forum of discussion about water resources management needed some more connections from the global scale to regional and national scales to be established, so that national water planning debate about the future of water resources could be more directly linked to global discussions about water management. This is how they decided to design the World Water Vision exercise (Cosgrove and Rijsberman 2000).

Once the strategic diagnosis has been made, and once the objectives for the intervention are set, the methodological reflexion about the design of the foresight intervention can begin. The change in perspectives that we propose also changes the point of view on the range of possible approaches for intervention, putting in question the fact that foresight methods might be standard.

Frequently, designing a foresight procedure directly consists of designing a participatory scenario exercise. In some cases, methodological questions can nearly be reduced to technical questions about the composition of the group of people involved in the procedure or the enhancement of the simulation model used to help people envision the consequences in the future of the decisions they might take. Of course, more complex methodological questions consist of designing the most relevant selection and arrangement of tools (participatory techniques, scenarios, models). We would argue that the participatory scenario exercise itself is only one of many very diverse possible approaches to deal with a specific foresight intervention situation.

A very good example of the fact that in some situations the participatory scenario workshop is not the most relevant approach for intervention is the case of California's water planning debate in the 1990s. A water specialist from a private research institute, Peter Gleick, also very present at the international scale, was confronted at the scale of the federal state of California with a water planning debate dominated by the water planning administration and its technical hydraulical representation of future water scarcity problems, thus leaving no room for the discussion of alternative modes of development for water uses (imagining changes in crops and cropping patterns in irrigation, modifying in the long run the features of urbanism), not to talk about the absence of ecosystems and their resilience in such a representation. In order to open the debate to other representations of future water scarcity problems and solutions, and to allow better access by other groups of interest, Gleick decided to make a counter proposal in the debate, building a possible scenario of California waters' future where aquatic ecosystems are preserved, groundwater is not overexploited and water demand is stabilised (Gleick 1995), and showing that such a scenario was plausible without assuming greater or faster changes than in the 20 years before. The specificity of his strategy was that he chose not to gather all stakeholders and the administration around the same table to try and build from scratch a shared vision, but that he rather chose to pave the way to a really pluralistic debate about the future of California's water resources by showing how to build an alternative scenario, even if not in a participatory way, and how to discuss its plausibility. By doing so, he took part in the debate and proposed to take into account new dimensions of the problem of water scarcity, and at the same time he made it possible for other groups to enter a debate that was formerly closed around the ability to build a water plan only from the technical engineering perspective. Such an intervention, where the participatory aspect is secondary with respect to the importance of the contents of the scenario that is being built and the way it is put into discussion, seems very close to the kind of strategies that are illustrated in 'Scenarios: the art of strategic conversation' (Van der Heijden 1996).

This example illustrates that every foresight intervention situation is specific and that the range of possible approaches for intervention is wide open and not restricted to standardised procedures or packages of tools and procedures. The foresight specialist might quite well build a scenario on his own and put it to discussion, in order to exemplify what are the rules for discussion about the quality of a scenario, or in order to propose a reframing of the analysis to other dimensions of the future of the theme studied. He might also very well only propose a procedure enabling the involvement of new emerging groups in the debate, without intervening in the contents of the discussions about the future. Alternatively, he might intervene in a combined way on the procedure and in the contents of the debate, for which the combinations of possible procedures and possible methods to represent future evolutions is close to infinite, or at least not limited to a few standardised approaches.

In this second section, we have tried to illustrate that the conceptual framework that we propose:

- enables the status of a foresight intervention to be reconsidered,
- puts the stress more on the initial diagnosis of the situation of intervention rather that on the implementation of universal procedures, in order to set context specific objectives,
- and also enables the ranges of possible approaches for intervention to be reopened, so that methodological dilemmas, choices and questions can really be adressed with respect to the specific objectives of intervention, rather than being reduced to technical questions concerning the implementation of a universal method.

Perspectives for the evaluation of a foresight intervention

If the initial diagnosis has been explicit, leading to the identification of specific objectives of intervention, for instance the reopening of the water planning debate to non-technical representations of the future and non-technical stakeholders in the case of California, a specific foresight intervention approach will have been chosen, and a specific intervention strategy designed and implemented.

The evaluation of the quality and performance of the foresight intervention can then clearly be formulated, in terms of incremental change with respect to the initial situation: did the implemented intervention strategy lead to the changes and the improvements expected in the future-oriented debate?

To measure the improvements in the changes of a future-oriented debate, we have to consider both dimensions:

- How did we manage to improve the quality of the representations of the future that are mobilised in the debate?
- How did we manage to improve the accessibility of the discussion to new emerging stakeholders and to new points of view?

Based on these two structuring questions, a series of criteria can then be used for both dimensions. If we look at what De Jouvenel (1964) presents as the criteria of a good representation of the long-term future, the quality of the representations of the future can be judged on the internal consistency of the future images of the systems that are envisioned, on the quality of the causal path represented, on the transparency of the assumptions that are made and on the ability of the representation of future evolution to be put into discussion. Concerning the quality of the forum of discussion, we can think of the criteria proposed by the sociology of science in order to evaluate the quality of hybrid fora and their ability to enrole emerging groups in the discussion about scientific and technological controversies (Callon, Lascoumes, and Barthe 2001). Criteria for example can be taken from the following list proposed by Callon, Lascoumes, and Barthe (2001) for the evaluation of transformations of 'hybrid fora':

(1) Evaluation of the results of the intervention in the forum
  (i) on the deliberative dimension:
    - diversity and pluralism of the points of view represented in the forum
    - openness and access to the forum by emerging groups
    - precocity of the involvement of emerging groups
    - diversity and autonomy of groups involved
  (ii) on the cognitive dimension:
    - quality of the representations discussed
    - rigour of the discussion
    - continuity of the discussion

(2) Evaluation of the means mobilised
  - What were the guarantees of the possibilities of access to the forum for emerging groups or points of view?
  - What resources (time, money, education) were made available for emerging groups?
  - Transparency and publicity of the debate, traceability of precedent interventions and representations put into discussion
  - Clarity and publicity of the rules of discussion.

This list of criteria should be adapted and it is not clear that an evaluation using these criteria would be less dependent on long-term monitoring than other criteria, but they do propose some precise identification of features of the forum. They might therefore enable a precise strategic diagnosis of the initial state of the future-oriented debate: the evaluation of incremental change in one of these criteria might be more feasible and also more useful in the perspective of enhancing the performance of foresight processes. This example does not intend to propose the only possible criteria, but rather to illustrate that operational links could be made between the practice of foresight and the theoretical frameworks from the field of science and technology studies.

This example, taken from one specific reference in the field of the sociology of science, linked to actor network theory, could be completed by some other references from the same field, or from the field of management sciences and collective learning (for example, the presentation of foresight as a collective learning process by Bootz (2001)). It might be of interest to take note that in the perspective taken here, foresight essentially refers to the action of managing changes and transformations of the future-oriented debate, which is similar to a collective strategy elaboration process.

The intention of this paper was to illustrate that by considering foresight as an intervention aimed at an incremental change in an existing future-oriented debate it enables at the same time much more specific and precise evaluation questions to be asked, and also gives access to a variety of possible criteria in analogous fields of interest.

Conclusion

This paper considers foresight as the activity of intervening in a future-oriented debate or a collective strategy elaboration process in order to improve the quality of the debate. A foresight exercise in this perspective can only be considered as a moment of possible inflexion in the general transformations of these collective deliberative processes. It is useful to consider that collective long-term strategies cannot be built only with one foresight exercise, but that they might emerge from a longer lasting collective learning process, if the future-oriented debate is pluralistic enough and if the criteria for discussing the quality of the representations of the future are explicit enough.

The paper illustrates that such a conceptual framework can make it easier to design, implement and evaluate foresight activities that would be much more context specific. This conceptual framework makes it easier for the field of research on foresight methodologies to connect with analogous fields like the sociology of science, or the field of strategy in management sciences.

References

Alcamo, J., and T. Henrichs. 2000. World water in 2025: global modelling and scenario analysis for the World Commission on Water for the 21st Century. World Water Series, Report no. 2, Kassel.

Berkhout, F. 2006. Normative expectations in systems innovation. *Technology Analysis & Strategic Management* 18, no. 3–4: 299–311.

Bootz, J.P. 2001. Prospective et apprentissage organisationnel. Travaux et recherches de prospective, no. 13, Futuribles International, Paris.

Callon, M., P. Lascoumes, and Y. Barthe. 2001. *Agir dans un monde incertain: essai sur la démocratie technique.* Paris: Seuil.

Clark, P., C. Booth, M. Rowlinson, S. Procter, and A.Delahaye. 2007. Project Hindsight: exploring necessity and possibility in cycles of structuration and co-evolution. *Technology Analysis & Strategic Management* 19, no. 1: 69–81.

Cosgrove, W., and F. Rijsberman. 2000. *World water vision: making water everybody's business.* London: Earthscan/World Water Council.

De Jouvenel, B. 1964. *L'art de la conjecture.* Monaco: Editions du Rocher.

Falkenmark, M., and G. Lindh. 1974. How can we cope with the water resources situation by the year 2050? *Ambio* 3, no. 3–4: 114–122.

Gleick, P. 1995. *California water 2020: a sustainable vision.* Oakland, CA: Pacific Institute for studies in Development, Environment and Security.

Godet, M. 1989. Effective strategic management the prospective approach. *Technology Analysis & Strategic Management* 1, no. 1: 45–56.

Godet, M. 1991. *De l'anticipation à l'action: manuel de prospective et de stratégie.* Paris: Dunod.

Godet, M. 1998. La boîte à outils de prospective stratégique. *Cahiers du LIPS CNAM – Futuribles international* 5.

Margat, J. 1992. Water in the Mediterranean Basin, situation and prospects. Blue Plan Report no. 6, Sophia Antipolis.

Martin, B.R. 1995. Foresight in science and technology. *Technology Analysis & Strategic Management* 7, no. 2: 139–168.

Ohlsson, L., and A.R. Turton. 1999. The turning of a screw: social resource scarcity as a bottleneck in adaptation to water scarcity. MEWREW Occasional Paper no. 19, SOAS Water Issues Study Group, University of London.

Reger, G. 2001. Technology foresight in companies: from an indicator to a network and process perspective. *Technology Analysis & Strategic Management* 13, no. 4 : 533–553.

Rijsberman, F. 2000. *World water scenarios – analysis.* Paris: UNESCO. http://www.worldwatercouncil.org/Vision/library.shtml.

Sreenath, S., and M. Mesarovic. 1997. Globesight sustainable development: water resources stress and scarcity: case study, Nile River problematique. Case Reserve University, Cleveland, OH.

Treyer, S. 2005. La disponibilité des ressources naturelles en eau comme facteur limitant du développement. Un débat prospectif à l'échelle mondiale. In *Etudier des écologies futures, un chantier ouvert pour les recherches prospectives environnementales,* ed. Laurent Mermet. Bruxelles: EcoPolis – Editions PIE – Peter Lang

Treyer, S. 2006. A quelle raréfaction de l'eau faut-il se préparer? Construire une intervention prospective au service de la planification pour les ressources en eau en Tunisie. PhD diss., ENGREF, Paris.

Van der Heijden, K. 1996. *Scenarios: the art of strategic conversation.* Chichester, UK: John Wiley.

Green technology foresight of high technology: a social shaping of technology approach to the analysis of hopes and hypes

Michael Søgaard Jørgensen and Ulrik Jørgensen

Department of Management Engineering, Technical University of Denmark, Kgs. Lyngby, Denmark

High tech visions play an important part in public technology policy and are often promoted through technology foresights. The article presents and analyses results from a green technology foresight of nano-, bio- and information- and communication technologies initiated by the Danish Environmental Protection Agency with the purpose of acquiring knowledge about the environmental potentials and risks related to the three areas of technology. The foresight was organised with a social shaping of technology (SST) approach to the field in order to cater for the complex relationship between societal demands, technology options, innovation dynamics and environmental impacts. The approach involved studying actor-networks, laboratory programmes and technology trajectories as well as deconstructing different stakeholders' high tech visions. The identified environmental potentials and risks related to the three areas of technology and recommendations for future governance of research, innovation and application areas are discussed.

Introduction

The world is facing large societal challenges as a result of environmental problems stemming from the large resource consumption in the industrialised countries and the increasing consumption in the newly industrialised countries, partly based on outsourcing of manufacturing from the industrialised countries. In this context many expectations are assigned to so-called high-tech areas like nano-, bio-, and information- and communication technologies (ICT). These are, for example, supposed to contribute to solving environmental problems. However, these expectations often build on rather linear and simplistic understandings of the properties of these technologies and their ability to deliver solutions. Changing the view on technological change from a linear understanding to a social shaping of technology approach (SST) to technological change has profound theoretical implications for the way future technological change in technology foresight is conceived. Instead of taking technology as an (external) driver of change, limited only by the ingenuity of mankind, technology is seen simultaneously as a driver and as the object being assigned tasks by different actors to be a part in solving societal problems. Technological change is a continuous process, where technology and social and environmental aspects are co-shaped by the priorities and focus in research, in innovation and in the applications of technology. In this article we discuss the experiences of employing a social shaping of technology approach to a green technology foresight of nanotechnology, biotechnology and ICT we were engaged in during 2004–2006 as researchers at Technical University of Denmark (Jørgensen et al. 2006). 'Green' was understood as a foresight study with the main focus on environmental aspects and impacts. Although great expectations were assigned to these technologies from the outset, the questions raised in the green technology foresight initiated by the Danish Environmental Protection Agency were still open-ended concerning the environmental potentials and risks. This created an opportunity to develop ways to deconstruct guiding visions of researchers and companies, focus on the different levels of maturity concerning innovative potentials of the areas, and ask questions of the complementary technologies needed to implement the high-tech visions in societal applications. It also helped identifying spaces where emerging and existing technologies can be addressed for governance of their potentials and risks with respect to the environment.

After an introduction to the background of the green technology foresight project the research design is presented, including those SST approaches which were applied in the foresight activities. Furthermore, the results from the green technology foresight project are discussed: (a) the identified environmental potentials and risks related to the three areas of technology, and (b) the proposals for future governance of research, innovation and societal application areas, as a prerequisite to forefront environmental potentials and risks in the development of high tech areas.

The aim and background of the green technology foresight

The aim of the green technology foresight was to develop recommendations for Danish strategies for integrated environmental and innovation policies in relation to the three areas of technology in order to reduce and prevent the future risks to environment and ensure focus on the environmental potentials in research, innovation and applications, and obtain environmentally related competitive advantages (Jørgensen et al. 2006).

The foresight was funded by the Danish Environmental Protection Agency (EPA) and managed by the Department of Manufacturing Engineering and Management (now Department of Management Engineering) at the Technical University of Denmark (DTU) and the System Analysis Department at Risoe National Laboratory (now also part of DTU). The focus of the foresight was argued with reference to the expectation that hi-tech-based products and systems could provide more sustainable solutions, including reduced chemical exposure without changing technological regime, such as moving from conventional farming towards organic farming, or from private cars towards public transport. The choice of EPA reflected the central role these technologies are assigned in research and innovation policies almost all over the world, but reflected also the environmental potentials and risks often highlighted in relation to these technologies (Anon. 2003).

Research design: green technology foresight as discursive activity

This paragraph explains how the foresight study was organised as discursive activity with focus on the co-shaping of technological change and environmental aspects. The SST approach to technological change (Bijker 1995; Russell and Williams 2005) and historical studies of technological development (Hughes 1987) have demonstrated the mutual influence of science and society on technological development and the interdependencies of technologies. Technology is seen as a bricolage (Latour 1999), a mixture of different elements, and technological change as a continuous process, where technologies and their societal impacts are co-shaped by a series of actions taken in research, innovation and application.

The SST approach and its combined focus on production and consumption of technology implies that environmental aspects and impacts cannot be assigned as properties to materials or processes *per se*, but are outcomes shaped during activities of research, innovation and application. Seemingly rather identical technologies can be applied and handled in very different ways and contexts resulting in very different environmental impacts. The following two examples of technologies analysed in the report from the green technology project illustrate this complexity of environmental impacts (Jørgensen et al. 2006). Few would probably disagree in the development and use of new systems of miniaturised sensors combining ICT and nanotechnology, organised as nets of sensors and providing possibilities for the measurement of concentrations of chemicals in nature. However, environmental NGOs and researchers emphasising cleaner technology would probably object if the sensors ended up being used for measurements of emissions for the purpose of maintaining the discharges just below a given emission level and thereby changing the focus away from prevention at the source. The use of organic material (biomass) for production of bioethanol might sound as an environmentally friendly strategy substituting hydrocarbons, but the actual environmental impact depends on the type of organic material (is it organic waste or plants grown especially for bioethanol) and the alternative use of organic material (e.g. maintaining the humus layer in soil or heating and electricity production in power plants, which also substitute fossil fuels).

The examples illustrate how environmental aspects of science and technology are shaped through the contexts of application including users and the interaction with other technologies and not just shaped in the processes of research and innovation. New, generic technologies have to be combined with already existing and new, complementary technologies to reach the level of practical application (Andersen and Jørgensen 1997; Freeman and Perez 1988; Wengenroth 1993). Consequently the specific strategies and conditions of application were included in the foresight, which led to the development of advice not only for research and innovation policies but also for the governance of sector policies (Jørgensen et al. 2006).

With the words of Brown, Rappert, and Webster (2000) we see the role of SST based foresight as to 'look at the future' rather than 'look into the future' and trying to be predictive. When leaving behind the idea of just outlining future developments, foresight studies inspired by SST can open for not only different descriptions of possible futures, but also the engagement of societal actors in the controversies around the paths of development to be preferred and the risks and benefits, which society at large will be forced to handle. Such controversies are necessary and productive (Elam and Bertilsson 2002; Norges forskningsråd 2005):

- From a *democratic* point of view: the citizens have to live with the consequences of the research and the products and since the public funds for research and innovation are limited the priority of some research and innovation strategies will limit the funds available for alternative strategies for achieving the same types of values.

- From a *pragmatic* point of view: citizens and NGOs can contribute with other perspectives than the researchers and other experts, due to other experiences and other values.

Analyses in the green technology foresight

Five different analytical approaches were employed in the green foresight study in order to collect, develop and analyse information about ongoing research, innovation processes and applications including plans and visions for future research, innovation and applications (Jørgensen et al. 2006):

(1) Analysis of *emerging applications of technologies* within the three areas of technology. Impacts of corporate practice and other stakeholder practices, structural conditions in existing and emerging value chains, and patterns in the environmental potentials and risks. This included surveys of the prerequisites for further dissemination and application of such technologies.
(2) Analysis of *present research and innovation*, emphasising the mechanisms of prioritisation in research and innovation, the role of existing knowledge regimes in research and innovation, and visions shaping and framing research and innovation, including the eventual role of environmental concerns.
(3) Organisation of *dialogue processes* among proponents of the three areas of technology and actors from environmentally important areas of product applications in Denmark followed by an analysis of how these actors envisage possibilities of applying technologies for environmental improvements within the three areas.
(4) Development of *scenarios for probable, future paths* of innovation and application within the three areas of technology and possible alternative development paths.
(5) Identification of *environmental potentials and risks related to the three areas of technology* followed by a comparison of these potentials and risks with societal discourses on environmental problems and discussions of whether there are better ways of solving the addressed environmental problems. Based on these analyses, recommendations for integrated environmental and innovation policies were developed.

The methods applied combined desk research based on literature, qualitative interviews with resource persons within the areas of technology and stakeholder workshops with participants from a number of stakeholder groups, including researchers, companies, governmental authorities and NGOs.

Theories inspiring the green technology foresight

This paragraph explains how theories within the SST approach have been applied in the foresight study in order to identify spaces and situations, where socio-technical change can be analysed, addressed and opened for political controversy. Actors and institutions were seen as undergoing mobilisation, displacement and reconfiguration (including the establishment of new actors and institutions), as an integral part of technological change (Clausen and Yoshinaka 2004). The following four fields of theory have been applied in the five analyses in the green technology foresight (Jørgensen et al. 2006):

(1) Research and development processes with researchers, artefacts, theories and visions, were seen as laboratory programmes and techno-economic networks in an actor-network perspective.
(2) Innovation processes of stabilisation and transitions in systems of innovation systems, including technological regimes and path creation/dependency were seen in the economics and theories of innovation perspective.
(3) Environmental assessment was organised as social and scientific processes by using methods from life cycle thinking and methods of dialogue-based environmental assessment.
(4) Science and technology as policy network processes involving many different stakeholders, aspects of innovation and regulation was understood within a theory of governance perspective.

These four sources of inspiration are detailed in the following paragraphs.

Actor-networks in innovation

The actor network theory (ANT) developed by Callon (1986), Latour (1992) and Law (1991) is useful in technology foresight for several reasons. An actor-network approach argues that a technology is not just working through a technical artefact, but as an emerging and increasingly stabilised network of associations between diverse material and non-material elements, such as artefacts, humans, texts, symbols and concepts. The notion 'actant' is used to encompass the communality of both the human and non-human and the material and non-material elements in an actor-network. Actor-network as a notion refers to the definition of actors through the network relations they are part of. The approach can be extended to follow the network-building strategies of actors and stresses the mutual constitution and transformation of these elements in the generation of agency, knowledge, institutions and power as effects of the network-building (Russell and Williams 2005).

In a foresight project an actor-network approach is useful in the analyses of both emerging and existing applications within an area of technology and of the priority mechanisms in research and innovation and the visions about future applications. The approach supports a focus on all the elements that seem to be necessary to make a technology either work or not work. In the foresight study this approach enabled, for example, a broader focus on a technology element such as nanoparticles, which included the supporting technologies needed to make this technology operational, as well as the need for, standards and measurement protocols for making the particles work in a technically and socially acceptable way, and the societal agendas they refer to (such as around competitiveness, health and environment). Thereby the approach supports a focus on those actants, which researchers and developers, sometimes explicitly but often unconsciously and implicitly, enroll into a network in order to make the technology become a working entity of a well defined kind and stabilised in such a way that it can be assigned certain properties (Jørgensen et al. 2006).

Laboratory programmes

The concept of laboratory programmes was used in the project to frame and identify how researchers organise the focus of their research and to demonstrate that research processes are not arbitrary and non-biased search processes. A laboratory programme is a sociological approach to research processes that applies the idea of actor-networks on the daily activities of scientists in the laboratory, understood as any kind of research environment. By applying this concept it is possible to identify what preconditions and assumptions that influence the choices and draw the attention of the researchers. The tools of the researcher are the assumptions, the instruments, the theories and not least the anticipated outcomes that constitute the laboratory setting. This implies that we have been tracing the implicit or explicit guiding visions of the researchers and seen these as very important for the framing of problems and interpretations and the practical repertoire of activities employed in the research process (Latour and Wolgar 1986).

Problems, which the researchers claim need to be solved, and the related solutions are seen as outcomes shaped in parallel during the research, when certain achievements are reached in the research process. What is taken into account as legitimate elements and included in the process as, for example, problems or parameters within the researchers' understanding and what is considered outside and difficult to approach are both shaped at the same time and in the same process. A laboratory programme will become more stable when instruments and theories are attached to the programme and alignment processes takes place, where physical objects and actors, for example, are given roles that mutually support the research activities and provide them with a legitimate perspective.

An example from the foresight project is the analysis of the discourses around genetically modified (GM) food and plants, which shows examples of such processes of co-shaping of problems and solutions, including the reframing of the activities within a different research perspective (Jørgensen et al. 2006). Originally GM researchers and companies pointed to pesticide resistant plants as an efficient agricultural strategy but after critique from environmental NGOs the use of GM plants was translated into an environmental strategy due to its claimed potential for reduced pesticide consumption. However, other researchers and environmental NGO's pointed to the risk of getting locked into a pesticide-dependent track and the risk of transferring genetic material coding for pesticide resistance to other related plants (Hansen 2000; Kuyek 2002). A similar example from the nanotechnology area seems to be the focus on applications of so-called nanosilver in product categories such as textiles, household equipment and paint as a solution to fight pathogenic bacteria and guaranteeing the consumer a clean household. Bacteria are regarded as an enemy to human beings, despite the fact that we could not exist as a species without the huge role played by bacteria in the human digestive system. The use of nanosilver has caused controversies about the risk of generating microorganisms that are immune to the well-known anti-microbial impact of silver, which is needed in the treatment of serious burns (see e.g. Nanotechnology Law Report 2007).

Techno-economic networks

The techno-economic networks (Callon 1991), which the interviewees (including researchers and companies) either are part of, or which they explicitly or implicitly anticipate will be developed in the future as part of possible future technology applications, were identified during the analysis of emerging applications and the priority mechanisms in research and innovation. In the analysis focus is on the dynamics between the past experience of the interviewee, the ongoing activities and the interviewee's thoughts about the future development. Furthermore, focus is on relations with existing technology paths and how these seem to influence the research and innovation or how the innovation paths and the companies and institutions shaping and reproducing them might be challenged or might be enrolled in certain visions for the future. An example in the foresight study was the analyses of the shaping of researchers' and developers' work with the future use of ICT for process optimisation, which showed the important role of a new interest from the

US FDA (Food and Drug Administration) for more flexible approaches to process regulation in the pharmaceutical industry, compared to the existing GMP approach (Good Manufacturing Practice), which demands the same process parameters from batch to batch. The new focus of the FDA seems to influence research and innovation within process modelling and process regulation directed towards the pharmaceutical industry (Jørgensen et al. 2006).

Changes in technological regimes and trajectories

When dealing with future technological development and its shaping, the SST approach stresses on the one hand the contingency and unpredictability of technological or rather socio-technical change and the plurality of actors involved, and on the other hand the structures and constraints shaping or restricting changes. Among the concepts used to describe regularity in technological changes in the foresight study were:

- Trajectories and paradigms (developed by Dosi 1982).
- Regimes as the combination of rules supporting a trajectory and guiding innovation (as defined by Russell and Williams 2005).

A technological trajectory is the pattern of problem solving activity within a given technological paradigm. Economic priorities, together with social and institutional frames, will operate as selective devices as new trajectories emerge. Dosi (1982) defines a technological paradigm as a pattern of solution of selected technological problems. In (Russell and Williams 2005) a regime is defined broader than a paradigm and includes furthermore elements such as scientific knowledge, engineering practices, technologies and skills. The regime reduces uncertainty and influences the search space in innovation processes. The criticisms of the focus on regularities in innovation processes argue that the extent of patterns easily is overstated and often are identified afterwards. However the stress on fluidity and contingency has raised concerns too, because the analyses mainly are complex descriptions without recognition of the patterns and continuities that can be observed and the social structures, which these patterns reflect.

Technological regimes were found in the area of ICT and also in the emerging applications of biotechnology. Within the field of ICT computer systems has been defining working conditions in industrial production planning as well as household consumption of media and communication. The overall rationalisation impact of this technology is well documented in industrial production while the impact on administrative work and on household consumption has been critical in environmental terms and led to the need to regulate electronics waste (Jørgensen et al. 2006). Within genetic modified crops the technological regime includes non-reproducible, pesticide resistant seeds as well as the pesticides that the seeds are made resistant to from the biotechnological companies, which builds dependency among the farmers of the biotech companies (Hansen 2000; Jørgensen et al. 2006).

Identifying visions and constructing possible future paths of research, innovation and application

The approaches of techno-economic networks and of regimes and trajectories support the technology foresight process in two ways (Jørgensen et al. 2006). First, in the analysis of *the emerging applications* of a technology it is necessary to understand the background for the breakthroughs, the dead ends, and so on, in the research and development activities. It was for example not enough to know whether it becomes possible to manufacture a certain type of sensor, but also whether this manufacturing is based on, for example, a certain type of equipment, material, co-operation with others and demand from clients, is important knowledge. This tells about path dependency and path creation in research and innovation (and thereby also the potential influence of certain equipment or clients in the future). It was also important to try to make the developers explain the technological systems around the applications such as necessary supply of energy and materials, standards and competencies, which are emerging or need to emerge, so that relevant life cycles and environmental aspects could be identified and analysed (Jørgensen et al. 2006).

Second, in the analysis of *research and innovation* it was important to understand the background and the prerequisites for the expectations the actors have. What is the role they are anticipating, e.g. nanoparticles will have (a certain behaviour in terms of reactivity or stability etc.) and who are expected to be the future users, and in which technological systems does this imply the nanoparticles will be integrated? What are scientific and technological breakthroughs considered as necessary in order to obtain a 'working' version of whatever component it might be? The answers to these questions make it is possible to develop a picture of the future research needs as seen by the actors and sketch elements in future life cycles as basis for life cycle assessments of environmental potentials and risks (Jørgensen et al. 2006).

Identification and assessment of environmental potentials and risks

A methodology for the assessment of environmental aspects within areas of technology was developed as an iterative process during the project. The assessment of the future environmental aspects and impacts were based on (Jørgensen et al. 2006):

- *Life cycle thinking*, which means assessments 'from cradle to grave'. Some assessments were focusing on present emerging applications and others on the information from researchers about possible future applications, which enabled the sketching of product chains for assessment of environmental aspects.
- *Systems approach*, which implies that not only single techniques are taken into consideration, but also other system elements (like infrastructures), which the techniques are parts of or dependent of, are included in the assessment.
- A broad, *dialogue-based understanding of 'environment'*. The understanding of relevant environmental aspects was shaped through, for example, studies of literature and dialogue at project workshops. Part of the approach was inspired by the approach of 'participatory life cycle assessment' (Bras-Klapwijk 1998), where the focus of life cycle assessments is discussed among the concerned and involved actors in order to increase the legitimacy of the assessment among the actors afterwards.

An assessment of the environmental aspects of the future development within the areas of technology is complicated owing to the many unknown elements within the future development in research and innovation and the different societal areas of application and consumption. This challenge in the environmental assessment was dealt with by focusing on the societal problems and discourses, which have been addressed in the past and recent development, and on the understanding of societal problems and discourses and application areas, which researchers and companies addressed, when interviewed about the focus of research and development.

Besides the life cycle and system perspectives, it is also of importance for the assessment of environmental impacts of a technological change whether it implies substitution of previous technologies (with, for example, less positive environmental impacts) and/or implies an increased consumption of products and materials. A framework developed by Berkhout and Hertin (2001) for assessments of the environmental impacts related to the use of ICT was adapted and combined with the life cycle and system perspectives (Jørgensen et al. 2006).

Identified environmental potentials and risks

The following paragraphs presents the findings from the green technology foresight within the three areas of technology in a Danish perspective, which means potentials and risks involving Danish stakeholders as either researchers, innovators and/or applicants of technologies within the three areas of technology (Jørgensen et al. 2006).

Findings related to ICT

ICT is an area of technology, where present and previous applications make it possible to analyse how ICT development and applications have co-shaped the environmental impacts, whereby it was possible to analyse whether and how environmental concerns are addressed and managed as part of ICT-related technology paths. The ICT sector in Denmark has a strong position in communications technology and pervasive computing and is one of the leading countries regarding public and private use of ICT. Environmental potentials and risks were analysed within five application fields: (a) development of the environmental knowledge base, (b) improved product and process design, (c) improved process regulation and control, (d) intelligent products and applications and (e) reduction of transport through changes in logistics and mobility needs.

The use of ICT-based tools and devices for data collection and processing of more data in more complex calculations demonstrated potentials for environmental advantages from future ICT applications. Some tools enable more overall resource efficiency in for example industry so that it is the specific aim of the application by the single company that determines whether and how environmental achievements are in focus. The integration of electronic components into products, so-called intelligent products or pervasive computing and intelligent applications, could imply environmental potential from automatic optimisation of the function of products such as operational feedback to the user, and digital product information about maintenance and reuse. The overall picture, however, was that the improvements will be related mostly to optimisations in energy and resource usage, rather than dramatic reductions in the overall environmental impacts.

Telework, e-business and logistics are ICT applications with strongly appraised implications, especially in relation to future transport. Telework might imply that regular transport related to commuting and shopping in certain hours is replaced by more differentiated transport needs. However, although only a limited amount of employees will be able to telework this could challenge the existing infrastructure of public transport and strengthen individual transport solutions. Mobile telework will probably be more widespread and offer the same facilities at comparable costs as those offered from the office and thereby enable an increase in business travel in conjunction with the ongoing globalisation of manufacturing and trade. Within freight transport, e-business and the concept of just-in-time production in industry could imply more transport due to the request for more frequent supply of small batches of materials and products. Logistic tools might optimise the amount of transport, but the impacts are not simply related to the use of ICT but as much a consequence of transport costs and policies.

The identified environmental potentials within the five fields of application play no significant role in the development and use of software and ICT equipment today. Achievements within these fields demand environmental regulation of the respective application areas, which can influence the priorities made by the users and the dominant driving forces in research and innovation.

The increasing amounts of electronic products, miniaturisation of products, pervasive computing and a more dispersed use of sensors and other devices have, in contrast to the idea of ICT being environmentally friendly, led to increasing problems with electronic waste. An efficient implementation of the EU directive about reduction of hazardous substances in electronic products (RoHS) is seen as necessary in order to achieve the anticipated substitution of some toxic materials in the future (Jørgensen et al. 2006).

Findings related to biotechnology

Positive visions of the environmental contributions of biotechnology developments have been prevalent since the 1980s but the environmental visions have historically not been the dominating drivers of innovation. Strong efficiency drivers especially for pharmaceutical and agricultural applications of new biotechnology have dominated the development. The environmental agenda has only recently come more to the fore with a number of reports, policy documents and discussions papers, which refer to a number of more specific biotechnology developments with environmental perspectives. The motivations for these developments are referred to as an increasing emphasis on problems in the chemical-using industries, on resource scarcities and on the need to 'clean up'. Together with environmental regulation, government priority setting and biotechnology regulation, these have been drivers of an increasing, but still small part of biotechnology, with the potential to address environmental issues.

In Denmark, enzyme technology is promoted as a key technology for realisation of environmental potentials of biotechnology. Reference is made to applications, where enzymes address and reduce the application of toxic chemicals and resource consumption, including energy consumption. Representatives in industry as well as researchers expect enzymes will contribute further by increasing efficiency in production and use of enzymes, by being applied within more industries and by further use in industries already using enzymes.

With regard to developments within bioethanol and biopolymers, environmental concerns have been key motivators for innovation, together with concerns for fossil fuel scarcity. The environmental assessments demonstrate, however, a need for further evaluation and debate of the perceived environmental potentials. Future use of biological resources has to address the issue that biological resources are of limited availability, and environmental claims must compete with the use of the same land and crops for food and for fuels in the energy sector power plants (Jørgensen et al. 2006).

Findings related to nanotechnology

The environmental aspects related to the nanotechnology area are yet very uncertain. There are knowledge gaps both about environmental risks and potentials, because of the very early stage of development. There is rising international and national concern about environmental and health risks related to nanoparticles, which questions the overall environmental potential of nanotechnology and is seen as an argument for urgent inquiry by environmental and health researchers and by some nano researchers. The foresight project showed that Danish nano innovation trajectories are still at an early and fluid stage of formation. The current phase of path creation seems crucial to the direction that nanotechnology is going to take. The analyses of the researcher visions and corporate visions showed that, despite frequent references to environmental potentials in the general debate on nanotechnologies, the problem solving activities and the emerging technological paths among Danish nano researchers and companies are only to some extent addressing environmental potentials. Environmental risks are also overlooked, although there is a rising concern about these among stakeholders involved with nanotechnology.

The project identified nano-related environmental potentials within different parts of the nanotechnology area, which may contribute to abatement and prevention of environmental effects in four ways: (1) tailored, efficient products; (2) new materials with properties, which enable reduced consumption of energy and other resources in the manufacturing and/or the use of these products and materials; (3) technology for renewable or more efficient energy systems, including fuel cells, solar cells and wind turbines; and (4) environmental remediation with more targeted treatment of pollutants and more efficient catalytic cleaning. Many of these areas are in the early stages of research and innovation, but by the 2020s they may offer possibilities to abate and prevent environmental problems related to energy supply and transport, resource use, use of chemicals and treatment of waste and wastewater. Most of the identified environmental potentials are, however, not likely to be pursued in the current research and innovation trajectories in Denmark and need policy action to be realised. Furthermore, the existing environmental assessment procedures are not adequate for measuring and handling materials at the nanoscale (Jørgensen et al. 2006).

Recommendations for future governance of the three areas of technology

The project anticipated at the outset that technological change and environmental impacts would, only in a few special cases, be directly linked to the development of an area of technology and the materials or processes part of this development. The environmental impacts of, for example, the chemical industry is not only shaped by the development of ICT-based tools for process design and regulation, but also by innovation in process equipment and by the national and local governmental regulation of the wastes and emissions of industrial facilities. The studies of the three areas of technology confirmed this conceptual anticipation within all three areas despite their differences, including the different degrees of application. Based on the identified future environmental potentials and risks and inspired by earlier analyses of and experiences with governance of science and technology, three core policy perspectives were developed in the foresight project (Jørgensen et al. 2006):

(1) *Guiding research* to include environmental perspectives, including policy options for assessing research strategies and potential outcomes, creating visions and objectives for areas of research and prioritising the research to be supported.
(2) *Focusing innovative activities* on the combination of technologies within specific fields of application to establish specific strategic innovation policies supporting the creation of new paths of development and bridging the gap between research and market diffusion.
(3) *Regulating technology applications* through the regulation of driving forces and institutional frames determining the development of consumption areas and based on coordination of policies between different policy domains.

Within all three policy perspectives no single actor like for example a national government can set the course, as the results of the policies are dependent on the networks of actors cooperating and the controversies and resulting stabilising strategies. This is reflected in the *environmental governance* approach to policy developed as a fourth cross-cutting perspective. A deliberate focus on environmental aspects in research and innovation policies needs to establish a framework of social interactions focussing on the environmental concerns of different stakeholders and addressing the uncertainties in relation to both potentials and risks related to new technologies (Gibbons et al. 1994; Jørgensen et al. 2006).

Environmental policy usually addresses mature technologies looking for substitution possibilities to remedy urgent environmental problems. It is quite another issue to address generic technologies, or even, as in the case of nanotechnology, an area of technology that only is beginning to materialise. Obviously the interventions and questions raised are different in the idea generation phase compared to the later experimental, early production or full commercial phases of innovation processes. The following paragraphs discuss the proposals for future governance of research, innovation and application related to the three high tech areas in general and specifically for each of the three areas.

Governance of research

An important question to address in future research policy is when and how to introduce policy measures in relation to research in order to obtain an increased focus on environmental potentials and risks. The following types of guidance were identified in the foresight study (Jørgensen et al. 2006):

(1) Visions for the anticipated societal role of an area of technology and the environmental outcomes as a means of shaping the research policies and also the rules of attention in the research community.
(2) Environmental screening of research proposals as a way of qualifying the decisions and priorities made in research funding.
(3) Guidelines for the organisation of environmental assessment of technology research as integrated and/or independent research activity.

Measure (1) is inspired by the need to challenge the utopian ideas of researchers and the often tacit prerequisites such visions are building on (Grove-White et al. 2004). Dialogue processes with broad participation of interested stakeholders, including organisations seldom invited into planning and management of research (e.g. consumer organisations, environmental organisations and trade unions) might be a way of enabling assessments of the social and technical prerequisites of technology visions and of the environmental aspects of research and innovation.

Measure (2), screening of research proposals, is only a valuable measure if the researchers are able and willing to address these issues and, at the same time, those evaluating the proposals agree in the importance of environmental aspects *and* are able to assess these aspects. Some of the earlier Danish experiences within food technology research with a request for applicants to address aspects of the impacts on environment, health and safety have not shown too promising results. Many researchers chose not to address the issues and the evaluators did not find the aspects important enough to let these aspects influence the funding decisions. The experience with the present demand for addressing socio-economic aspects in EU research applications has neither been too promising, as many of the technical domains only have addressed these issues rather superficially (Jørgensen et al. 2006).

Measure (3) is inspired by strategies developed during the years 2000–2005 in the UK and the Netherlands with environmental assessments in relation to nanotechnology research. An important aspect of the assessment of environmental aspects of technology concerns the research funding schemes and the organisation of the environmental research as integrated into or independent of the technology research. Environmental researchers who are part of a nanotechnology research group could, on the one hand, develop trust in the relations with the nanotechnology researchers and have enough proximity to the research subject to allow detailed assessments. On the other hand, integrated environmental research without independent funding could become too dependent of the nanotechnology researchers. Two European examples show how environmental concerns may be related to nanotechnology research. At the University of Cambridge a researcher, who was trained as a chemist before switching to the social sciences, was employed within a nano research group. His activities included development of methods for reflection on the social implications of the research and considerations about how non-scientists might be involved in deciding what the nanotechnology research should focus on. The challenge is to avoid making the social sciences the handmaiden of technological innovation (Turney 2005). The Dutch constructive technology assessment (CTA) (Schot and Rip 1997) activity within the national Dutch nanotechnology programme, NanoNed, is organised as an independent, but integrated programme. The CTA programme has its own projects, but in connection to activities within the NanoNed programme. The aim of the technology assessment is to understand and improve the interaction between science, technology and society (Technology Assessment 2006). The challenge is how the CTA research avoids becoming social marketing of nanotechnology.

In relation to ICT the green technology foresight identified a need for more research about the actual and the potential roles of ICT-based technologies in the shaping of less polluting and resource consuming use patterns in order to develop more socio-technically advanced paradigms for ICT-based technologies. In relation to biotechnology more research about the environmental aspects of biotechnology seems to be one of the prerequisites for future application of these technologies. The key barrier to focus on environmental potentials related to nanotechnology is lacking awareness and knowledge of nano-related environmental potentials and business potentials (Jørgensen et al. 2006).

Governance of innovation

The integration of environmental aspects into strategic innovation policies aims at creating new paths of development and bringing new technologies to real life test. The identified limited focus on environmental potentials in innovation within all three areas of technology shows the need for measures, which can ensure environmental focus in innovation processes (Jørgensen et al. 2006). The results of this kind of measures should be the creation of new paths for technological development by supporting the critical and highly uncertain first steps of bringing ideas for potential environmental benefits from the research stage to real prototypes and tests. The approaches applied in the proposals are inspired by Danish experiences within cleaner technology (Andersen and Jørgensen 1997), cleaner products (Remmen 2002), wind power (Karnøe and Garud 1998) and organic food (Jørgensen 2004). The Danish experiences demonstrate that it is possible to develop new, more sustainable development paths within an application area in competition and cooperation with existing technological regimes. Such path creation demands a combination of reshaping of existing institutions, competencies and regulatory mechanisms, and development of new institutions, competencies and regulatory mechanisms. The overall proposals for measures include (Jørgensen et al. 2006):

- Support for innovation combining technologies into products for specific fields of application whereby the environmental impacts better can be identified and realistic user conditions confront the technologies.

- Integration of environmental concerns into the innovation processes at the earlier stages of laboratory and prototype developments are important to assure that these aspects are part of the creation of development paths.
- Support for market development through combinations of regulation of potential application fields, support for demonstration projects, and network activities involving potential suppliers, customers, knowledge institutions and intermediaries.

In relation to ICT the foresight study showed the need for more focus on the potentials of and limits to intelligent products and applications as elements in environmental strategies. Furthermore, need for strategies to ensure focus on hazardous substances and materials and radiation in the development of products and components. In relation to biotechnology there is need for development of enzymes with environmental potentials for a broader variety of industrial processes and a strategy for sustainable use of biomass as renewable resource. There is need for considerations about how the industrial up-take of nanoscience can be based on environmental potentials. A central barrier is lacking environmental competencies among Danish nano researchers, lacking nano competencies among environmental experts and in industry, and weak linkages among these groups (Jørgensen et al. 2006).

Governance of areas of application

If mature and market-introduced technology applications with environmental potentials are not realised under present market, production and user regimes more stringent regulatory policies could provide a difference. A sector or product domain approach may be needed instead of a technology approach (Jørgensen et al. 2006). Some findings from application areas identified in the analyses of the three areas of technology are presented in the following.

ICT-based process control is used in many industries and this kind of control could be a way of reducing environmental impacts if the process control is focused on reduction of resource consumption and emissions. However, stronger governmental regulation of industry in terms of demands to emissions combined with increased prices on resources and support for competence development is necessary to encourage industry to reduce the emissions through better process control and also wider use of enzymes (Jørgensen et al. 2006). The amount of transport of persons and freight is dependent on a number of socio-economic changes that have very little to do with the development of ICT equipment for transport and logistics. The amount of private transport depends more on the costs of housing in areas near the workplace of households and prices on public transport and fuel. Hybrid cars with engine and batteries with ICT-based control of the vehicle could give significant improvements in fuel efficiency. However, as a result of the present innovation strategies in the car industry and the tax system on vehicles and fuel, the market penetration of these types of cars is very modest (Jørgensen et al. 2006).

In relation to nanotechnology, regulation of applications is not yet a key instrument considering that most of the identified environmental potentials are pre-commercial. In specific cases, e.g. new types of energy efficient lighting, regulation of the application of nanotechnology could be feasible, but at a later stage (Jørgensen et al. 2006).

Conclusions

Visions of environmental potentials within all three areas of technology exist, but tend to be based on quite simplistic assigned properties to the technologies as idealised entities. At the same time there is only little recognition of the need to regulate research, innovation and application areas to reach a situation where the environmental potentials are critically assessed and where environmental aims become a core element in the priorities of research and innovation. In research and innovation policies the decoupled environmental characteristics assigned to the generic technologies seem to prevail. A number of more specific applications have been prioritised and argued based on their environmental potentials, though lacking the governance support to ensure the anticipated environmental potentials.

The foresight study has demonstrated that it is possible to identify future environmental aspects and impacts related to different paths of development within the studied areas of technology. The social shaping of technology approach, including the environmental assessments, applied in the foresight study enabled analyses of the shaping of present and previous societal applications of a mature area of technology such as ICT and of a somewhat mature area of technology such as biotechnology. Dynamics co-shaping research, innovation and applications were identified and enabled deconstruction of environmental hypes and hopes. The mapping and deconstruction of researcher visions about future application areas related to all three areas of technology, including the emerging nanotechnology area, have furthermore enabled identification and deconstruction of possible future environmental aspects and impacts. In relation to all three areas proposals for more advanced future governance of research, innovation and applications were developed.

Even though the developed approach to green technology foresight has been able to create a more complex understanding of the relation between technology, society and environmental aspects and impacts, the project has shown a need to focus more on sector transitions and the involved applications of technology instead of taking generic technologies as the outset. In such foresight studies the emphasis should be on the dynamics, which shape the application areas in order to avoid the translation of societal problems into simple technical solutions. Add-on technologies such as transport logistics and nanosilver may be faster to apply to societal problems, compared to changes in transport needs or intensive agriculture and its impact on bacteria in food. However, simplistic technical strategies will not abate and prevent the environmental problems and may even create new problems, such as the bacteria resistance in the case of nanosilver mentioned earlier, or divert the awareness of the complexity of the problems and the need for coherent and integrated solutions.

References

Andersen, M.S., and U. Jørgensen. 1997. The support for cleaner technologies: results, diffusion and future activities – an evaluation of the support programmes 1987–1992. Environmental Review no. 14, Danish Environmental Protection Agency, Copenhagen.

Anon. 2003. Baggrundsnotat til sigtelinie 2 Innovation og spredning af miljøteknologier – 2. ansøgningsrunde 2003: 2.1 Grønt teknologisk fremsyn om design af miljøvenlige produkter og materialer (In Danish) [Background paper for line 2 innovation and diffusion of environmental technologies – 2nd application round 2003: 2.1 Green technology foresight on design of environmentally friendly products and materials].

Berkhout, F., and J. Hertin. 2001. Impacts of information and communication technologies on environmental sustainability: speculations and evidence. Report to the OECD, Science and Technology Policy Research Unit, University of Sussex, Brighton.

Bijker, W.E. 1995. *Of bicycles, bakelites and bulbs: towards a theory of sociotechnical change.* Cambridge, MA: MIT Press.

Bras-Klapwijk, B. 1998. Adjusting life cycle assessment methodology for use in public policy discourse. Delft University of Technology.

Brown, N., B. Rappert, and A. Webster, eds. 2000. *Contested futures: a sociology of prospective technoscience.* Aldershot: Ashgate.

Callon, M. 1986. Some elements in a sociology of translation: domestication of the scallops and fishermen of St Brieuc Bay. In *Power, action and belief: a new sociology of knowledge?* ed. J. Law, 196–233. London: Routledge & Paul Kegan.

Callon, M. 1991.Techno-economic networks and irreversibility. In *A sociology of monsters: essays on power, technology and domination*, ed. J. Law. London: Routledge.

Clausen, C., and Y. Yoshinaka. 2004. Social shaping of technology in TA and HTA. *Poiesis & Praxis* 2, no. 2: 221–246.

Dosi, G. 1982. Technological paradigms and technological trajectories: a suggested interpretation of the determinants and directions of technological change. *Research Policy* 11: 147–162.

Elam, M., and M. Bertilsson. 2002. Consuming, engaging and confronting science: the emerging dimensions of scientific citizenship, STAGE (Science, Technology and Governance in Europe) Discussion Paper 1, Gothenburg University and Copenhagen University.

Freeman, C., and C. Perez. 1988. Japan: a new national system of innovation? In *Technical change and economic theory*, ed. G. Dosi, C. Freeman, R.R. Nelson, G. Silverberg and L. Soete. London: Pinter.

Gibbons, M., C. Limoges, H. Nowotny, S. Schwartzman, P. Scott, Peter, and M. Trow. 1994. *The new production of knowledge*. Thousand Oaks, CA: Sage.

Grove-White, R., M. Kearnes, P. Miller, P. Macnaghten, J. Wilsdon, and B. Wynne. 2004. Bio-to-nano? Learning the lessons, interrogating the comparison. (Amended version.). Lancaster University and Demos.

Hansen, A.G. 2000. Danish development of genetically engineered herbicide resistant beets and the role of environmental controversies. Danish country account of the growing of genetically modified plants and their use. Report for the EU project 'Consumer, environmental and sectoral networks in the societal management of food technology'. Unpublished report. Department of Technology and Social Sciences, Technical University of Denmark, Lyngby.

Hughes, T.P. 1987. The evolution of large technological systems. In *The social construction of technological systems*, ed. W.E. Bijker, T.P. Hughes and T. Pinch, 51–82 Cambridge, MA: The MIT Press.

Jørgensen, M.S. 2004. Policy integration in the shaping of organic food as strategy in the Danish food sector – what happened to the environmental concern as driving force. Paper presented at a Conference on 'The Human Dimension of Global Environmental Change: Greening of Policies – Policy Integration and Interlinkages', 3–4 December 2004, at Free University of Berlin, Germany.

Jørgensen, M.S., M.M. Andersen, A.G. Hansen, H. Wenzel, T.T. Pedersen, U. Jørgensen, M. Falch, B. Rasmussen, S.I. Olsen, and O. Willum. 2006. *Green technology foresight about environmentally friendly products and material – the challenges from nanotechnology, biotechnology and ICT.* Working Report no. 34, Danish Environmental Protection Agency, Copenhagen. http://www.mst.dk

Karnøe, P., and P.R. Garud. 1998. Path creation and dependence in the Danish wind turbine field. Papers in Organization no. 26, Institute of Organization and Industrial Sociology, Copenhagen Business School, Copenhagen.

Kuyek, D. 2002. Genetically modified crops in Africa: implications for small farmers. Brief written for GRAIN (Genetic Resource Genetic Resources Action International), Spain.

Latour, B. 1999. *Pandora's hope – essays on the reality of science studies.* Cambridge, MA: Harvard University Press.

Latour, B. 1992. Where are the missing masses? The sociology of a few mundane artefacts. In *Shaping technology, building society*, ed. W.E. Bijker and J. Law, 225–258. Cambridge, MA: MIT Press.

Latour, B., and S. Woolgar. 1986. *Laboratory life: the construction of scientific facts.* Princeton, NJ: Princeton University Press.

Law, J., ed. 1991. *A sociology of monsters: essays on power, technology and domination.* London: Routledge.

Nanotechnology Law Report. 2007. http://www.nanolawreport.com/tags/nanosilver/ (accessed 6 January 2008).

Norges forskningsråd. 2005. Nanoteknologier og nye materialer: Helse, miljø, etikk og samfunn. Nasjonale forsknings- og kompetansebehov [Nanotechnologies and new materials: health, environment, ethics and society. National research and competence needs]. Norges forskningsråd, Oslo.

Remmen, A. 2002. Greening of Danish industry – changes in concepts and policies. In *Technology studies and sustainable development*, ed. A. Jamison and H. Rohracher, 171–198. München: Profil.

Russell, S., and R. Williams. 2005. Social shaping of technology: frameworks, findings and implications for policy with glossary shaping concepts. In *Shaping technology, guiding policy: concepts, spaces and tools*, ed. K.H. Sørensen and R. Williams, 37–133. Cheltenham: Edward Elgar.

Schot, J., and A. Rip. 1997. The past and future of constructive technology assessment. *Technological Forecasting and Social Change* 54, no. 2: 251–268.

Technology Assessment. 2006. About technology assessment in the Dutch NanoNed programme. http://www.nanoned.nl/TA/Technology+Assessment.htm (accessed 4 December 2006).

Turney, J. 2005. Before you build that nanobot. *The Times Higher Education Supplement*, 28 January 2005. http://www.nanoandsociety.com/news/documents/THES%20-%2005.pdf (accessed 4 December 2006).

Wengenroth, U. 1993. How they won the market: electric motors in competition with steam engines, 1890–1925. In *Technological competitiveness. Contemporary and historical perspectives on the electrical, electronics, and computer industries*, ed. W. Aspray, 270–286. Piscataway, NY: IEEE Press.

Management of foresight portfolio: analysis of modular foresight projects at contract research organisation[1]

Totti Könnölä[a], Toni Ahlqvist[b], Annele Eerola[b], Sirkku Kivisaari[b] and Raija Koivisto[b]

[a] *Institute for Prospective Technological Studies, Joint Research Centre, European Commission, Edificio Expo, C/Inca Garcilaso, 3, 41092 Seville, Spain;* [b] *VTT Technical Research Centre of Finland, P.O. Box 1000 02044 VTT, Finland*

While the expansion of foresight scope towards systemic processes and societal considerations has provided significant opportunities for learning and synchronised action between different business units and/or policy fields, it may also have caused digression and ambiguity in the practice and theory of the management of foresight processes. This is true, in particular, in contract research organisations that have faced major challenges to reorganise their foresight activities as part of the changes in their innovation practices. The paper examines the exercises and consequent responses to this shift at VTT Technical Research Centre of Finland. The paper develops and applies a coherent classification framework for foresight exercises. The framework provides practical support for the design and management of foresight projects, as well as supports the overall management of the portfolio of different kinds of foresight activities. The findings also point out the need for the modular process design that helps adjusting foresight exercises in different contexts.

1. Introduction

The gradual paradigm shift in the innovation research and policy from linear to systemic innovation models has challenged also the conventional technocratic technology-driven forecasting practices and called for new participatory and systemic foresight approaches (Smits and Kuhlmann 2004). In the 1980s, publicly funded foresight activities were commonly seen as an instrument for assisting in the development of priorities for research and development (R&D) resource allocation (Irvine and Martin 1984). Since then, stakeholder participation and networking have been regarded as increasingly important elements of foresight activities for 'wiring up' the multi-layered innovation systems both in public (Martin and Johnston 1999) and private sector (e.g. Salmenkaita and Salo 2004). Reports from foresight projects, in turn, have emphasised the importance of common vision-building as a step towards the synchronisation of the innovation system (Cuhls 2003).

In these developments, the locus of foresight activities has tended to shift from positive and rational technology-focused approaches towards the recognition of broader concerns that encompass the entire innovation system, including its environmental, social and economic perspectives. The High Level Expert Group appointed by the European Commission crystallised these trends by defining foresight as follows (European Commission 2002): 'A systematic, participatory, future intelligence gathering and medium-to-long-term vision-building process aimed at present-day decisions and mobilising joint action'.

While the expansion of foresight scope has provided significant opportunities for learning and synchronised action between different business units and/or policy fields, it may also have caused digression and ambiguity in the practice and theory of the management of foresight processes (Könnölä 2006). This is true, in particular, in contract research organisations that have faced major challenges when reorganising their foresight activities as part of the changes in their innovation practices. Indeed, Linstone (2007) has noted that separate corporate research laboratories are disappearing and/or transforming to market-driven contract research entities. This reflects the paradigm change in the R&D functions from the basic science and technology push driven innovation processes to the systemic innovations that emerge close to the market (Smith 2000).

The paper examines the recent experiences and consequent responses to this shift in the foresight activities at VTT Technical Research Centre of Finland. VTT is the largest contract research organisation in the Northern Europe, providing high-end technology solutions and innovation services. VTT has actively contributed to the development of national and international foresight practices. Towards this end VTT has coordinated and participated in a number of foresight processes aiming to provide improved contextual understanding of future developments strengthened networking and crystallised visions for joint action (for foresight objectives, see e.g. Könnölä, Brummer, and Salo 2007).

At VTT, as in many other contract research organisations, foresight activities are considered increasingly important throughout the R&D value chain. This implies a continuous stream of different types of foresight activities that also bring in understanding of how such activities together support the foresight objectives in the wider innovation system. Managing a larger set of foresight activities (or portfolio) and designing and managing individual, even strikingly different foresight projects requires profound understanding of the varying nature of the foresight tasks and expectations. For this purpose the paper first develops a conceptual framework for the classification of foresight projects. The framework is then applied to examine a portfolio of foresight projects, in which a foresight expert team at VTT – responsible of coordinating foresights projects involving experts from different VTT units and external stakeholder organisations – has recently been either a coordinator or a participant. The classification framework is expected to support the coherent and effective management of individual projects as well as the strategic management of the portfolio of foresight projects. It also clarifies the role of foresight activities within the organisation and in the whole innovation system in general. The findings of the analysis also point out the need for flexible modular design in the management of foresight projects.

The remainder of this paper is organised as follows. Section 2 defines and discusses the key design dimensions to be considered in the planning and management of foresight projects. Section 3 elaborates a classification framework for foresight projects making use of the four key dimensions identified. The classification framework is then applied to the analysis of a number of foresight processes that VTT has recently been engaged in. Section 4 reflects the findings from the empirical analysis and in particular examines the needs and reasons for modular design of foresight projects. Section 5 concludes the paper by discussing the contribution of the classification framework for the management of foresight portfolio and for the modular design and by pointing out interesting areas for further research.

2. Key dimensions of foresight project design

The systemic understanding of innovation processes has challenged conventional technology driven forecasting practices and called for new participatory foresight approaches that address also the consideration of diverse perspectives, formation of shared knowledge and examination of alternative futures. Foresight activities are also increasingly seen as crucial functions in order to prepare for the future – not only to identify the promising technological pathways but also to engage relevant stakeholders and create common visions and action plans (Eerola and Jørgensen 2002, 2008; Smits and Kuhlmann 2004; Ahlqvist 2005; Dannemand Andersen et al. 2007; Könnölä, Brummer, and Salo 2007). Furthermore, foresight processes can often be seen as a pertinent design phase for the creation of new value networks that are based on novel combinations of technologies, organisational partnerships and institutional arrangements. The design and management of such processes are prone to face major challenges in responding to the diverse expectations of the client(s) and other stakeholders.

Thus, the design of foresight activities can benefit from structured approaches which help to identify the expectations and challenges concerning the management of the process and the final outcomes. Towards this end we will discuss and define the key design dimensions in the management of foresight processes. The key design dimensions including *outcomes chosen future perspectives, management* and *stakeholder engagement* are first defined and elaborated and then applied to the analysis of selected foresight processes that the foresight experts of VTT have recently been engaged in.

2.1. *Informative vs instrumental outcomes*

Outcomes consist of outputs, results and impacts of the foresight exercise: outputs refer to the products and services, both tangible and intangible ones; results in turn refer to the advantage (or disadvantage) that the beneficiaries obtain soon after the end of their participation with the foresight; and impacts refer to consequences affecting beneficiaries during and after the project.

For the purpose of defining the key design dimensions, the nature of the outcomes of foresight activities can be divided in informative and instrumental outcomes. *Informative outcomes* refer to the use of foresight to improve the general understanding of present and future challenges of the innovation system and its parts. The informative outcomes do not refer to the expectations that foresight activity would necessarily lead to specific actions directly, although informative outcomes may increase the preparedness to act in some unspecified future situations (for example making it easier to recognise emerging risks and discover new windows of opportunities). *Instrumental outcomes, in turn,* refer not only to informative outcomes but also to the use of foresight to support the specific foreseen decision-making situation, for example related to resource allocation or the formation of strategic partnerships/joint actions.

Foresight processes not only elaborate on but also produce shared new knowledge for the stakeholders. New knowledge in the foresight project is essentially produced through collaborative work which benefits the strategic thinking and work of each participant. A useful framework for analysing such processes is the SECI model describing the dynamics of shared knowledge creation (Nonaka 1994; Nonaka and Takeuchi 1995; Eerola and Väyrynen 2002; Eerola and Jørgensen 2002; Eerola and Jørgensen 2008). In the SECI model, shared knowledge creation is envisaged as a spiral process in which tacit and explicit knowledge as well as the different modes of knowledge conversion, i.e. socialisation, externalisation, combination and internalisation, play a central role. 'Shared knowledge creation' refers here to *generation of new knowledge* that the stakeholders can share without necessarily agreeing on the exact meaning when applying it to specific problems.

2.2. *Consensual vs diverse future perspectives*

The nature of future perspectives needs to be addressed in order to define the ways and methods by which the project develops an understanding of the future. Foresight activities often focus on the production of *consensual future perspectives* that refer to the creation of common understanding on priorities, relevant collaborative networks and future actions. For example, the taxonomy of Barré (2002) reflects this trend by distinguishing between objectives for (i) setting scientific and technological priorities, (ii) developing the connectivity and efficiency of the innovation system and (iii) creating a shared awareness of future technologies. Alternatively, foresight activities can be seen as useful tools to identify *diverse future perspectives* in view of understanding diverse ideas, opinions and perspectives in priority-setting, identifying and fostering alternative and competing coalitions and value networks as well as exploring alternative futures and generating rivalling visions (Könnölä, Brummer, and Salo 2007). The value of communication of diverse perspectives and their inclusion in the decision-making have also been commonly recognised in the field of risk analysis (e.g. Koivisto, Reunanen, and Nissilä 1997; Koivisto, Törmäkangas, and Kauppinen 2002) and in societal embedding of innovations (Kivisaari, Lovio, and Väyrynen 2004).

In more specific terms, diversity can be linked to widely discussed foresight objectives (i.e. priority-setting, networking and common vision building) as follows (Könnölä, Brummer, and Salo 2007):

- Priority-setting supports the identification of common future actions and the efficient allocation of resources (Irvine and Martin 1984), but may decrease the diversity of options that could challenge conventional approaches and dominant designs (e.g. Arthur 1989) and escape from techno-institutional lock-ins (David 1985; Arthur 1994; Jacobsson and Johnson 2000; Unruh 2000). Here, foresight can generate ideas on alternatives and recognise diverse perspectives in priority setting (Salo, Gustafsson, and Ramanathan 2003; Keenan 2003).
- Networking enhances the connectivity of the innovation system and can improve its performance (Lundvall 1992; Martin and Johnston 1999), but may lead to the excessive strengthening of *existing* networks (see, e.g. Grabher and Stark 1997) creating path-dependencies and locking out alternative technological options (Unruh 2000). Thus, foresight should contribute also to the creative restructuring and even the destruction of lock-in conditions by engaging different stakeholders in the proactive generation of rivalling visions for competing coalitions based on different value networks with different architectures, configurations features and standards (Tushman and O'Reilly 1997: Könnölä, Salo, and Brummer to appear).
- Building a consensual vision of the future reduces uncertainties and helps synchronise the strategies and joint actions of different stakeholders (e.g. Cuhls 2003), but may lead to conservative and abstract results (Keenan 2003) to the effect that existing path-dependencies are further strengthened. General abstractions cannot form the basis of actions especially if responsibilities are not clearly identified (Salmenkaita and Salo 2004). The search for a consensual vision of the future should therefore be complemented with – or even replaced by – the exploration of alternative futures and respective techno–institutional arrangements (Könnölä, Unruh, and Carrillo-Hermosilla 2007).

The attempt to accommodate diverse perspectives on the future is also central to the methods for the scanning of weak signals. Ansoff (1975) defined weak signals as 'imprecise early indications about impending significant events'. Later on this definition has been expanded to accommodate additional characteristics, such as new, surprising, uncertain, irrational, not credible, difficult to track down, related to a substantial time lag before maturing and becoming mainstream (Coffman 1997; Harris and Zeisler 2002; Mendonça et al. 2004).

2.3. *From fixed to autonomous management*

Foresight process can be managed with different kinds of approaches often driven by diverse expectations laid on the project. Those in charge of foresight process are likely to benefit from the sharp definition of their role and approach in the management of the foresight process. This makes it easier to design the process in a coherent way and to communicate the responsibilities of different stakeholders so that there are also good conditions for carrying out a high-quality foresight process. For this purpose, two extreme approaches of classification can be identified. *Fixed management* can be characterised as a centralised approach in which co-ordinators fix the scope and methods of the exercise at the outset and control the process. This is often the case, for example, in Delphi exercises (Helmer 1983; Cuhls, Blind, and Grupp 2002). *Autonomous management*, in turn, refers to the process intermediated by the co-ordinators who facilitate an autonomous and evolving participant-led continuum of meetings and other activities. This may be the case for example in expert panel work (Salo, Könnölä, and Hjelt 2004).

The creation of new and shared, knowledge is challenging, in particular, when the people participating in the foresight process have heterogeneous backgrounds. This occurs when various interest groups (industry, academia, government, non-governmental organisations (NGOs), etc.) and different geographical areas (countries, regions, etc.) are engaging in the foresight process. Therefore, special attention must be paid to the organisation of the process and to the appropriate use of formal tools and procedures. In these circumstances the SECI model helps to understand the dynamics of the knowledge conversion and also serves as a tool for designing well-functioning autonomous foresight exercises emphasising collective learning (Eerola et al. 2004; Eerola and Jørgensen 2008).

2.4. *From exclusive to extensive stakeholder engagement*

The participatory aspect of the foresight process can be characterised in terms of the extensiveness of the stakeholder engagement, referring to the set of qualitative and quantitative factors including the number of stakeholders involved, openness to participate (whom are invited/allowed to participate) and diversity of stakeholders involved (diverse disciplines, policy and industrial sectors, NGOs, etc.). Thus, *extensive stakeholder engagement* refers to the approach in which the actual number of participants is high. The stakeholder participation is encouraged and open for all the interested stakeholders. Many kinds of stakeholders are invited to participate in the process. *Exclusive stakeholder engagement*, in turn, means that stakeholder participation is limited and is, thus, not open for all the stakeholders interested. The exclusive stakeholder engagement can provide opportunities for intensive stakeholder communication and, for example, dealing with sensitive issues related to intellectual property rights and competitiveness.

Building on Barré (2002) and Van der Meulen, De Wilt, and Rutten (2003), Salo, Könnölä, and Hjelt (2004) have distinguished three levels of stakeholder intensity of engagement with respective objectives:

(1) *Low engagement*: Stakeholders exchange ideas and perceptions on future challenges and comment on foresight deliverables, thus contributing inputs to the exercise. This, however, does not necessarily lead to notable changes in their perceptions.
(2) *Medium engagement*: Stakeholders engage in collaborative learning processes and proactive development of innovative options, to the effect that the perceptions of individual stakeholders are shaped by these processes.
(3) *High engagement*: Stakeholders are intensively involved in the collaborative management of the foresight exercise and also assume responsibilities in contributing to the development and realisation of jointly approved action plans.

In the implementation of stakeholder engagement it can be difficult to pursue both extensive and high engagement, because high engagement requires continuous and transparent processes of learning (Cruickshank and Susskind 1987). Such processes are likely to become time-consuming and expensive when the number of foresight panel participants increases (Hjelt et al. 2001; Van der Meulen, De Wilt, and Rutten 2003). However, without a sufficient number and diversity of participants the activity may fail to produce innovative results or to reach sufficient media attention or difficulties may be encountered during the up-take of results. Voss and Kemp (2006) discuss a parallel problem in reflexive governance. The problem consists of the contradicting requirements of opening up (towards the consideration of an extensive set of stakeholder perspectives) and closing down (towards focused analyses and intensive engagement).

In the subsequent sections the above discussed design dimensions of foresight management are used to characterise a number of foresight activities that the VTT horizontal foresight expert team has either coordinated or been participating in. The positioning of individual projects in the framework clarifies the methodological decisions and the rationales of stakeholder engagement. Once the projects are positioned in the framework they provide an overview of the whole portfolio of foresight projects that supports building the holistic view of the activities.

3. Foresight experiences at a contract research organisation

This section builds on the key dimensions of foresight project design discussed in Section 2 (see also Box 1). The classification framework is developed and applied to study some of the foresight activities that the foresight experts at VTT have recently been engaged in. These activities represent different regional organisational and sectoral contexts. Thereby, the section intends to create improved understanding on the methodological choices made during these projects and clarify how different foresight dimensions are linked to one another and to methodological choices.

The defined conceptual dichotomies of the foresight dimensions provide a useful structure for the analysis assuming, of course, that foresight activities consist of identifiable elements for the classification. In practice, foresight activities often consist of some elements of both sides of these dichotomies, and altogether they form the combination of a case specific process design. This is manifested in the summary of selected VTT foresight activities in Table 1.

Foresight projects can be further classified according to the design dimensions. When the dimensions of outcomes (informative vs instrumental) and future perspectives (consensual vs diverse) correspond with the horizontal and vertical lines, the projects (described in Table 1) can be positioned in four different quadrants (consensual and informative; consensual and instrumental; diverse and informative and diverse and instrumental) (Figure 1).

In parallel, the projects can also be positioned in view of process-oriented dimensions. When the process management (autonomous vs fixed) and stakeholder engagement (extensive vs exclusive)

- **Instrumental vs informative outcomes**
  - ***Informative outcomes*** refer to the use of foresight to improve the understanding of present and future challenges of the innovation system and its parts. Informative outcomes are not expected to lead to specific actions, although they can improve the preparedness to act in some unspecified forthcoming situations.
  - ***Instrumental outcomes*** refer in addition to informative outcomes also to the use of foresight to support the specific foreseen decision-making situation, for example, the situation may be related to resource allocation or the formation of strategic partnerships/joint actions.
- **Consensual vs diverse future perspectives**
  - ***Consensual future perspectives*** refer to the creation of common understanding on priorities, relevant collaborative networks and future actions.
  - ***Diverse future perspectives*** refer to understanding diverse ideas, opinions and perspectives in priority-setting, identifying and fostering alternative and competing coalitions and value networks as well as exploring alternative futures and generating rivalling visions.
- **Fixed vs autonomous management**
  - ***Fixed management*** can be characterised as centralised approach in which co-ordinators fix the scope and methods of the exercise at the outset and control the process. This is often the case in Delphi exercises.
  - ***Autonomous management***, in turn, refers to the process intermediated by the co-ordinators, who facilitate an autonomous and evolving participant-led continuum of meetings and other activities. This maybe the case in a expert panel work.
- **Extensive vs exclusive stakeholder engagement**
  - ***Extensive stakeholder engagement*** refers to the approach in which the actual number of participants is high, the stakeholder participation is encouraged and open for all the interested stakeholders, and many kinds of stakeholders are invited to participate in the process.
  - ***Exclusive stakeholder engagement*** means that stakeholder participation is limited and thus not open for all the stakeholders interested.

Box 1. The design dimensions of foresight management (own elaboration).

dimensions are considered to correspond with the horizontal and vertical axes, they produce together a coordinate system (see Figure 2). Here, the horizontal axis represents the qualitative continuum from fixed to autonomous management, and the vertical axis the continuum from extensive to exclusive stakeholder engagement.

Further on, if this coordinate system is positioned to each quadrant of Figure 1, the foresight projects can be positioned in the coordinates to provide detailed information on the nature of the outcomes and process of each project (Figure 3). Hence, once the project is in one of the four quadrants (according to consensual vs diverse and informative vs instrumental) the exact position of the project can be defined in the coordinates (from fixed to autonomous and from exclusive to extensive).

Positioning the projects in the quadrants, and in the coordinate systems within the quadrants, provides important insight for further analysis of their characteristics and methodological choices of the projects. Subsequently, we discuss the projects in each quadrant and discuss their positioning in the coordinate systems.

Table 1. VTT foresight projects in view of the design dimensions of foresight management.

Project	Outcomes	Future perspectives	Management	Stakeholder engagement
VTT Technology Futures Forum[1]	*Informative* Inform stakeholders on VTT foresight activities To foster and provide forum for future-oriented thinking and dialogue among VTT stakeholders	*Diverse* Present foresight results at VTT and among its stakeholders offering visions on alternative futures	*Autonomous* Participant empowerment Future-oriented workshops on selected themes facilitated by VTT experts	*Exclusive* Paid memberships for VTT clients interested in foresight partnership (planned option for deeper involvement) *Extensive* Seminar and workshop participation open and free for all stakeholders
VTT Water Research Roadmap 2006[2]	*Informative* Create common understanding on future challenges and VTT expertise *Instrumental* Support the formation of common R&D actions within VTT	*Consensual* The identification of key action areas for VTT water research and their priority-setting *Diverse* Inclusion of alternative views on priority-setting	*Fixed* Structured questionnaire Defined agenda for workshops Structured priority-setting *Autonomous* Flexible use of methods in working groups	*Exclusive* VTT expert engagement in steering group and workshops to enable intensive communication *Extensive* Networking among VTT experts on water related R&D through questionnaires and co-writing
VTT Security Roadmap[3]	*Instrumental* Present ideas for new VTT R&D Programmes on security technologies *Informative* To foster and provide forum for future-oriented thinking and dialogue among VTT researchers	*Consensual* Different themes presented in the same framework Shared understandings were searched for in order to find the key developments *Diverse* A variety of views and opinions were considered and debated during the process	*Fixed* Structured questionnaire Defined agenda for workshops Structured priority-setting *Autonomous* There was still a significant degree of freedom to adapt to the perceived needs during the process	*Exclusive* Participation by invitation only *Extensive* The participants were from different parts of VTT

Finnish Material Technologies Roadmap[4]	*Instrumental* Present ideas for new R&D Programmes on materials technologies *Informative* To foster and provide forum for materials technology development and dialogue among Finnish stakeholders	*Consensual* Different themes presented in the same framework Shared understandings were searched for in order to find the key developments *Diverse* A variety of views and opinions were considered and debated during the process	*Fixed* Structured questionnaire Defined agenda for workshops Structured priority-setting *Autonomous* There was still a significant degree of freedom to adapt to the perceived needs during the process	*Exclusive* Participation by invitation only *Extensive* The participants were from research institutes, industry associations and public organisations throughout Finland
Nordic ICT Foresight[5]	*Instrumental* Action proposals and policy recommendations *Informative* Evaluations of key ICT applications Nordic scenario set in context of ICT development Scenario-based visionary roadmaps Building views of the Nordic potentials in ICT development among key actors	*Consensual* Identification of ICT applications with development potentials in Nordic region Future-oriented elaboration of factors affecting the Nordic business and development environment in ICT *Diverse* Linking the perspectives of alternative scenarios to the research process	*Fixed* Structured discussion and generation of new ideas in the workshops *Autonomous* Creative brainstorming and ideation in the different workshops	*Exclusive* Intensive stakeholder engagement in core group *Extensive* Co-operative idea and concept creation among stakeholders from different Nordic Nordic organisations and firms; networking
Nordic H2 Energy Foresight[6]	*Instrumental* A clear action plan for the Nordic key actors – without a direct link to any decision process *Informative* Awareness raising and deepening the overall understanding of the entire value chain (hydrogen production, storage, distribution, stationary hydrogen uses and hydrogen uses in transport)	*Consensual* Shared understandings were searched for in order to be able to give action recommendations for the Nordic key actors *Diverse* A variety of views and opinions were considered and debated during the process	*Fixed* The overall design of the process was determined already when planning the project The model and modelling techniques in use guided the data gathering of the system analysis part *Autonomous* There was still a significant degree of freedom to adapt to the perceived needs during the process	*Extensive* The participants were from research institutes, industry, associations and public organisations of the five Nordic countries

(*Continued*)

Table 1. Continued.

Project	Outcomes	Future perspectives	Management	Stakeholder engagement
EU: Nanoroadmap[7]	*Instrumental* Clear guidance and recommendations for the new Framework Programmes were expected by the EC/DG Research *Informative* Present the key development paths and visions of nanotechnologies on energy applications	*Consensual* Shared understandings were searched for in order to find the key developments *Diverse* Different themes presented with the same framework A variety of views and opinions were considered and debated during the process	*Fixed* Structured questionnaire Defined agenda for workshops and structured priority-setting *Autonomous* There was still a significant degree of freedom to adapt to the perceived needs during the process	*Exclusive* Participation by invitation only *Extensive* The participants were from research institutes industry associations and public organisations Both EU and non-EU views were collected
EU: Emerging S&T Priorities in Public Research Policies[8]	*Instrumental* Clear guidance and recommendations for the new Framework Programmes were expected by the EC/DG Research *Informative* In addition to compilation of action recommendations, the process covered identification of new emerging technologies, their prioritisation, country-specific and regional analyses of socio-economic S&T environments, and construction of alternative action scenarios	*Consensual* Shared understandings were searched for between the project group and the client; For the purpose, it was also important to get a balanced European picture of the various topics *Diverse* A variety of views and opinions were considered during the process; Three European scenarios were produced outgoing from different sets of assumptions	*Fixed* The process consisted of extensive excel-sheet enquiries, systematic PESTE-type SEEP-analyses of socio-economic S&T environments, structured stakeholder interviews and interactive workshops together with the client's representatives *Autonomous* There was still a significant degree of freedom in the analytic back-office work and in compilation of action scenarios and recommendations	*Exclusive* Direct participation limited to European S&T experts some key persons from DG Research+the foresight experts of the project group *Extensive* In total, a considerable number of scientific and technical stakeholders from various European countries + US and Japan were consulted; Representatives of DG Research participated in the intermediate workshops and commented the intermediate reports
EU: HyWaysII in Finland[9]	*Informative* As part of HyWaysII integrated project to produce policy relevant information for the Commission and Member states	*Consensual* National level common vision on hydrogen demand up to 2050 *Diverse* Formulation of alternative hydrogen-based energy systems	*Fixed* Energy system modelling Stakeholder analysis *Autonomous* National workshops for the identification of national target levels	*Exclusive* Selected energy experts invited to the workshops; Intensive back-office work *Extensive* Energy stakeholders from different sectors and organisations

EU: IRRIIS scenario work	*Informative* Identification of emerging safety and security issues in an EU project to ensure the safety of critical infrastructures	*Consensual* A project level consensus on the future developments *Diverse* Different scenarios for the future development	*Fixed* A fixed procedure and selected methods *Autonomous* The content of the work resulted from an autonomous work among the stakeholders; The experience of stakeholders "overwrote" the methodological rigidity in some points	*Exclusive* The work was carried out among the project partners *Extensive* The results were tested against available expertise outside the project consortium
EU: ForSociety Dialogue on Transnational Infrastructures[10]	*Informative* To produce information for EU decision making and trans-national projects in the EU level	*Consensual* Consensus among the invited expert group, included also sharing of different views on the topics	*Fixed* Pre-planned process *Autonomous* Some room for creative panel work	*Exclusive* Invited participants only
EU: ForSociety Future Dialogue on Healthy Nutrition[11]	*Informative* To build a roadmap for research policy institutions; To contribute to national, EU and ETP discussions on future research and innovation	*Diverse* Creation of alternative scenarios for the future development	*Fixed* Pre-planned process *Autonomous* Some room for creative panel work	*Exclusive* Invited participants only

Notes: [1]http://www.vtt.fi/proj/tff/?lang=en; [2]Unpublished; [3]http://www.vtt.fi/inf/pdf/tiedotteet/2007/T2368.pdf; [4]http://www.teknologiateollisuus.fi; [5]http://www.vtt.fi/inf/pdf/publications/2007/P653.pdf; [6]http://www.h2foresight.info; [7]http://www.nanoroadmap.it/; [8]http://ec.europa.eu/research/foresight/pdf/21960.pdf, http://www.efmn.info/kb/efmn-brief42.pdf; [9]http://www.hyways.de/; [10]http://www.eranet-forsociety.net/ForSociety/resources/future_dialogues/; [11]http://www.eranet-forsociety.net/ForSociety/resources/future_dialogues/nutrition.html?dbX_sid=f82615f5016.

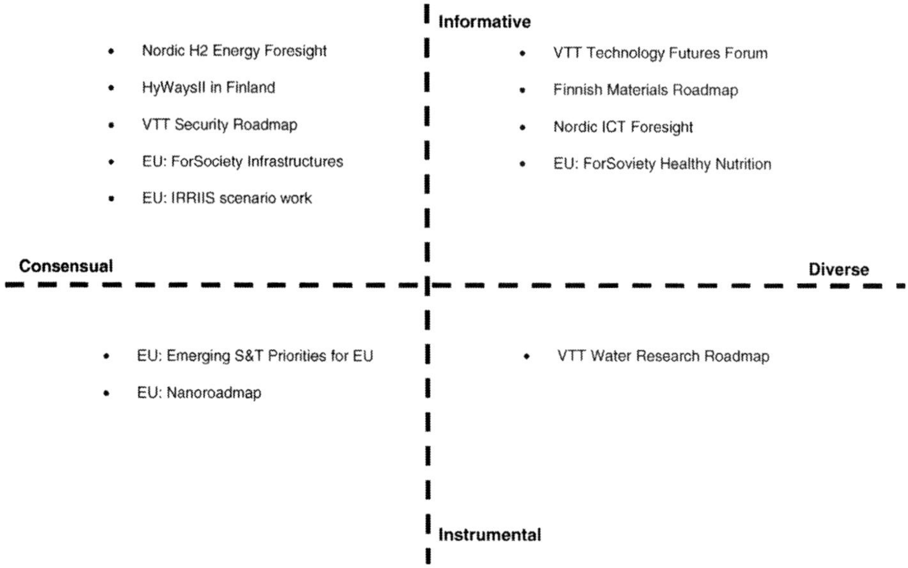

Figure 1. Foresight projects positioned in view of the dimensions of outcomes (informative vs instrumental) and future perspectives (consensual vs diverse).

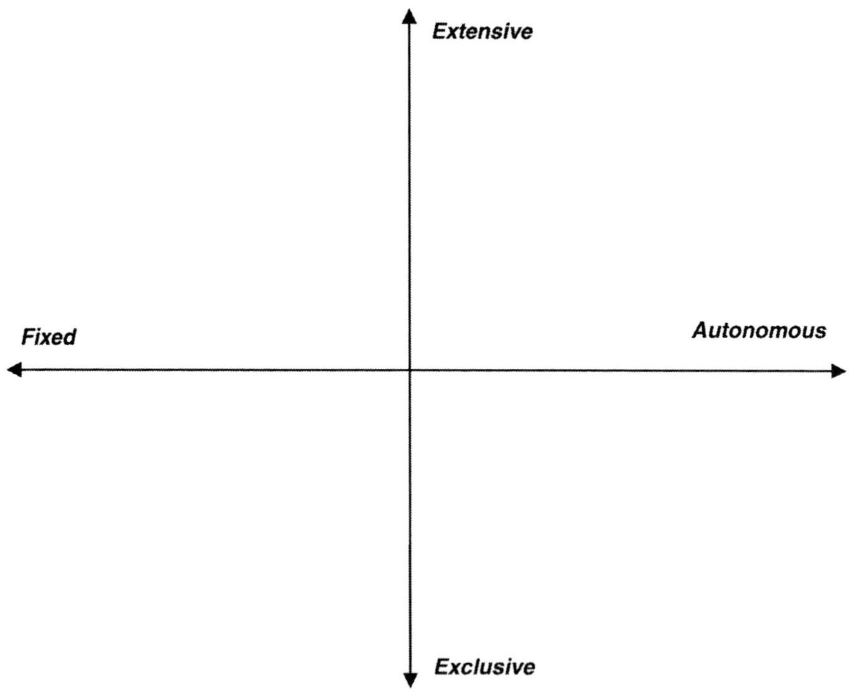

Figure 2. Process management (from fixed to autonomous) and stakeholder engagement (from exclusive to extensive) dimensions in a coordinate system.

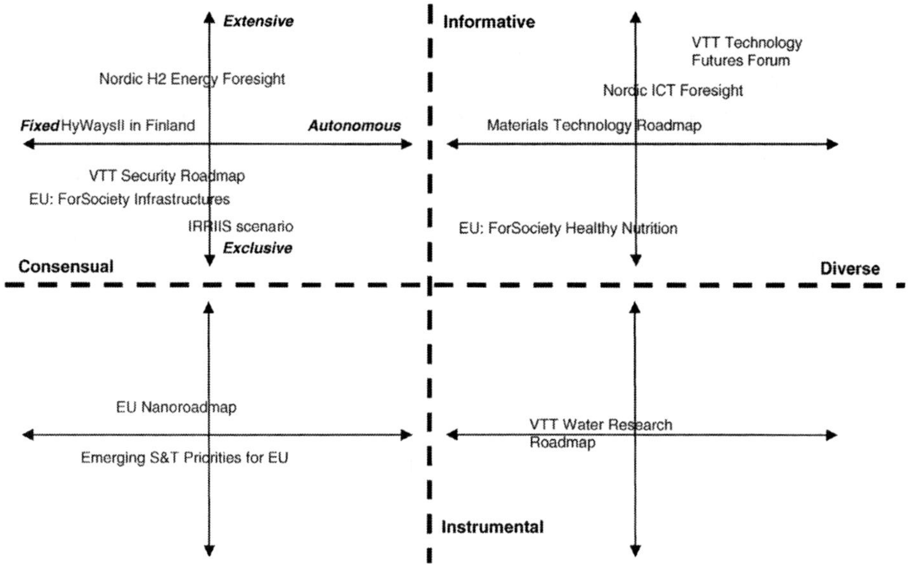

Figure 3. Foresight projects positioned in view of the dimensions of outcomes (informative vs instrumental), future perspectives (consensual vs diverse) and in the coordinate system of stakeholder engagement (extensive vs exclusive) and management (autonomous vs fixed).

3.1. *Consensual and informative processes*

Consensual, informative processes create common understanding on priorities, relevant collaborative networks and/or future actions. They are expected to improve the understanding of present and future challenges of the innovation system and its parts. However, specific short-term actions are not necessarily expected after the projects. This setting relieves the participants partly from claiming value and from the pressures of policy-making and lobbying. This also supports learning and common vision-building among adversary parties.

Two hydrogen-related foresight projects included visioning and modelling of future energy systems and they were characterised with fixed or partly fixed management, consensual perspectives and rather extensive stakeholder engagement. In the *Nordic H2 Energy Foresight* the major challenge was to create shared understandings between different stakeholder groups representing five different countries. This required getting the right balance between quantitative and qualitative approaches. Much of the efforts were directed towards the creation of a common language and understanding between technical experts, modelling people and foresight experts, and the various groups of stakeholders (industry, research institutes, public organisations and associations) in the five Nordic countries. The project applied the combination of interactive workshops, modelling, analytic back-office work and a small-scale Delphi-type enquiry. A quantitative model of the Nordic hydrogen energy system was built during the project. The stakeholders who participated in the foresight process provided also the inputs for the model.

HyWaysII in Finland, in turn, was conducted as a part of the European integrated project that explored the alternative scenarios of hydrogen demand up till 2050 in selected Member states. Market allocation modelling (MARKAL) with the focus on the demand side of the development of energy systems and the key changes and actor mapping methodology (KCAM) enabled the production of relevant inputs for European analysis but meant a strict structure and focus for backcasting and forecasting workshops. This created few opportunities for general discussion on the role of hydrogen solutions in Finland, which might have been even more productive for national hydrogen developments.

The consensual and rather fixed approach was applied also in the less extensive stakeholder process for *VTT Security Roadmap*. This process was challenged to address, on the one hand, the expectations on linking different units and operational project plans and, on the other hand, the strategic needs to construct roadmaps and identifying VTT priorities. Towards this end, a questionnaire was circulated on future technologies, challenges, opportunities and VTT expertise. Furthermore, workshops were organised for brainstorming, roadmapping and priority-setting, aiming at consensual statements on the future and corresponding R&D needs.

The *IRRIIS* scenario work was a more exclusive consensual and informative process on the future developments in energy and communication technology and their interactions. It was expected that the project results would describe the future scenarios in detail. This was considered very challenging by the participants who examined diverse uncertainties in such scenarios. This challenge was dealt with in the brainstorming workshops, intensive e-mail communication, commenting and co-writing.

The foresight exercise within the framework of the European Union (EU), *ForSociety Project – Future Dialogues on Transnational Infrastructures*, represented a joint European expert work with rather exclusive participation. The major challenge was to form small expert groups representative enough to cover diverse stakeholder perspectives for the pre-planned panel work in the workshops. These results were further complemented with experts' commentaries.

From the methodological and management viewpoint, *HywaysII in Finland* was carried out as a part of the European energy system modelling project, which required a rigid structure for workshops and the consensual fixed process design. The *ForSociety* infrastructures dialogue process, in turn, was guided by a general-level fixed process, but a creative autonomous stakeholder panel work offered still lots of freedom for invited participants to express their opinions to form consensual statements on the future. While informative processes supported visionary work producing consensual understanding on possible future directions the participation of decision-makers was scant (an aspect revealed by examining the participation dimension). In the *Nordic H2 Energy Foresight* and *HyWays II in Finland* specific efforts were made to engage policy-makers but with limited immediate success. This may be partly due to the initial positioning of the projects as informative rather than instrumental, and not considered as policy-making processes (Könnölä, Unruh, and Carrillo-Hermosilla 2007). However, indirect and diffuse policy links during and after such projects may be influential in the long run.

3.2. *Consensual and instrumental processes*

Consensual and instrumental processes create common understanding on priorities, networks and/or future actions as well as support the specific foreseen decision-making situation. Among decision-makers this is likely to lead to interests in the results. However, policy interests may also enter in the foresight process and create rigidities and difficulties that block new and fresh perspectives for change. The potential problem may be mitigated by ensuring extensive stakeholder participation through the diversity and high number of participants.

The EU foresight project on *Emerging S&T priorities in public research policies* engaged a large number of S&T (science and technology) stakeholders from industry, research institutes and funding organisations. The project applied both fixed and autonomous methods trying to balance with the transparency of the process, tight deadlines and budget and the genuine involvement of the client (DG Research). Towards this end, the methods used consisted of enquiries by email, PESTE-type (political, economic, social, technological and environmental issues) document analyses, structured stakeholder interviews and interactive workshops. In the *EU Nanoroadmap* the major challenges included how to get government officers, business leaders and academics engaged in the roadmap process and not just wait for the final results. This was pursued via a questionnaire and workshops on roadmapping and priority-setting.

The *IRRIIS* scenario work was a more exclusive consensual and instrumental process on the future developments in energy and communication technology and their interactions. It was expected that the project results would describe the future scenarios in detail. This was considered very challenging by the participants who examined diverse uncertainties in such scenarios. This challenge was dealt with in the brainstorming workshops, intensive e-mail communication, commenting and co-writing.

3.3. *Informative processes with diverse future perspectives*

Informative processes with diverse future perspectives take into account diverse ideas, opinions and perspectives in priority-setting, identifying and fostering alternative and competing coalitions and value networks as well as exploring alternative futures and generating rivalling visions. This relieves participants from the intensive search for consensus and direct support for decision-making. This 'released space' provides opportunities for creative thinking and the inclusion of diverse and alternative view-points that can challenge incumbent and path-dependent approaches hindering – especially radical – changes in the innovation system (Könnölä 2006).

Technology Futures Forum was planned as a free and open forum engaging VTT stakeholders in future-oriented thinking and dialogue. This enables the application of innovative methods enhancing creative thinking and formulation of possible, even radically different, futures. Extensive stakeholder engagement, including foresight practitioners, government officers, business leaders and academics, created a challenge on how to take different backgrounds in account in the seminars and workshops.

The Finnish material technologies roadmap supported the work of the Federation of Finnish Technology Industries and of the Finnish Funding Agency for Technology and Innovation (Tekes), and fostered networking of stakeholders from many different organisations with different backgrounds. In spite of the variety of views, the project managed to construct roadmaps with a rather fixed process structure and rigid management, including a questionnaire on future technologies, challenges, opportunities and Finnish competitive advantages, and workshops of brainstorming, roadmapping and priority-setting.

Dominating views, as well as institutional and political reasons, may sometimes downplay the consideration of diverse perspectives. This is particularly true in national and organisation-level foresight settings, whereas the presentation of diverse perspectives may be easier in cross-border foresight exercises where the participants are not aware of – and need not obey – organisational and national-level 'musts' and 'taboos' (Eerola et al. 2005). A good example of such a process is the European *ForSociety Future Dialogue on Healthy Nutrition*, in which invited stakeholders from different member states produced alternative future scenarios to support policy-making on the national and European level.

Another example is the *Nordic ICT Foresight*, which was designed using the similar modular process structure as in the *Nordic H2 Energy Foresight*. However, the project laid particular emphases on incorporating diverse perspectives in the vision, scenario-working and roadmapping. The major challenge in the process was to facilitate communication among stakeholders from various knowledge fields. ICT (information and communication technologies) is a broad theme, which made it difficult to form common understanding about the visions and highlighted ICT applications, especially when there were some personal-level changes in the stakeholder engagement along the way. Many subsequent phases – including desktop analyses, SWOT (strengths, weaknesses, opportunities and threats) workshops and questionnaires, scenario and vision work with facilitated brainstorming, clustering and assessments, and construction of structured socio-technical roadmaps – were needed to generate the resulting action proposals and policy recommendations.

3.4. *Instrumental processes with diverse future perspectives*

Instrumental processes with diverse future perspectives generate diverse ideas, opinions and perspectives, which support the specific foreseen decision-making situation or for the formation of strategic partnerships/joint actions. The *VTT Water Research Roadmap* was initially considered rather similar to *VTT Security Roadmap* to develop consensual and informative roadmaps. However, the key foci of the *VTT Water Research Roadmap* were the creative combination of wide-ranging water-related issues at VTT as well as the generation of new R&D initiatives. Questionnaires and mind-mapping and brainstorming workshops were used to engage VTT experts in roadmapping and formulation of innovation ideas. During the project, it appeared that there were diverse opinions on VTT technology expertise and future market opportunities in water research. This called for mapping these differences and identifying diverse project plans instead of producing consensus statements on common future priorities. The roadmaps were finally constructed through co-writing with different intensities of participant engagement.

4. Discussion

4.1. *Towards modular foresight design*

Section 3 introduced a general framework for the classification of foresight projects. It illustrated by examples from VTT foresight activities how different design dimensions focusing on outcomes, future perspectives, management and stakeholder involvement interact with each other. Conceptual work, theory-aided process design and systematic practices are, however, needed for successful integration of various qualitative and quantitative methods. In addition, it is important to consider case-specific conditions and flexible application of different methods, for instance, building on interactive workshop methods, online generation, elaboration and assessment of stakeholder ideas and opinions, and analytic back-office methods.

At the contract research organisation, foresight projects are carried out for a number of different clients and in different roles: as initiator, coordinator, cooperation partner and/or invited participant of a foresight process. This also means specific requirements with regard to the contract research organisation's preparedness to meet varying expectations of the clients and other stakeholders. Furthermore, the experiences on the VTT foresight processes summarised in Table 1 indicate that foresight management in a contract research organisation is prone to enjoy tensions produced by diverse policy and business driven expectations on the foresight process and the results. For example, the need to formulate relevant instrumental outcomes for decision-makers that lead to direct actions may create barriers to explore alternative futures, especially those with radical changes. On the other hand, the processes that are strictly informative may find it difficult to engage policy-makers and other decision-makers, who often could be valuable contributors in the foresight project. In many cases, the foresight projects seem to be best designed balancing between the different dimensions of foresight management.

In view of these empirical observations, we agree with Salo, Könnölä, and Hjelt (2004) that responsiveness – which refers to purposely instituted managerial mechanisms for making warranted mid-course adaptations to foresight objectives and implementation plans – should be regarded relevant in the management of foresight activities. In effect, responsiveness requires receptivity *vis-à-vis* the interests and expectations of participating stakeholders and flexibility in planning and implementation, achieved through the ability to envision and execute even radical changes in the foresight process (Salo, Könnölä, and Hjelt 2004). The need to adapt the foresight process to the requirements of the policy context and decision-making phase has been discussed by some other authors too (Eriksson and Weber 2006; Da Costa et al. 2008).

The responsiveness in the foresight management can be introduced through *modular foresight design*, inherent in the categorised foresight VTT projects and processes in Section 3. Modularity refers to process design where analogous sub-processes – or modules – can be enacted relatively independently from the other sub-processes (Könnölä et al. to appear). Thus, the modular foresight design refers to the planning and execution of foresight activities as modules that can be linked together to form a tailored processes for different needs. Modules themselves consist of certain standard elements (e.g. method) but also the modules can be somewhat tailored (e.g. target, scope, variety). As such, modular foresight design enhances the contract research organisation's capabilities for expected quick response to unique client needs. The modular foresight design can be useful also for addressing the foresight design dimensions:

- *Outcomes*: Accumulation of foresight knowledge for informative and instrumental outcomes can be enhanced via modular design that supports the commensurable and comparable process flows and results. This provides strong basis to create linkages between different issues and actors as well as detailed elaboration of outcomes in view of stakeholder needs. Moreover, the modular foresight design can make the assessment and comparison of the foresight processes and outcomes easier, more systematic and more transparent. For example, the SECI model referred to in Section 2 of this article allows the comparison of the foresight processes by setting them into a common conceptual framework. This setting makes the processes commensurable and makes room for critical reflection and further development.
- *Chosen future perspectives*: The modular design supports the application of different methods to address diverse perspectives on the future and to develop solid action plans. Hence, modularity helps balancing between the expectations on outcomes with regards to the emphasis on consensus or diverse perspectives.
- *Management*: The modular design allows the utilisation of the synergies between different foresight processes and makes it easier to link the results of sub-processes. Thus, it supports the attainment of scalability by enabling organising parallel processes. Within the modules even very rigid structures can be applied to assure that the objectives are met, while between the modules the coordinators can reflect and decide on even radical changes for forthcoming modules. On the other hand, the modularity can support the attainment of objectives through a rather rigid process structure keeping in mind that the results of one module should contribute to another. However, inside the modules, there can be lots of flexibility responding the stakeholder expectations and fostering creativity during the process.
- *Stakeholder engagement*: Modular design makes it easier for the different stakeholders to enter and contribute to the knowledge accumulation on the different project phases. Modular design thus supports reasonable division of work between foresight experts of different organisations

and between experts of different technical fields and backgrounds. Furthermore, the wide consultative processes and stakeholder engagement can be included without losing the control and strategic focus of the project.

At VTT, there seems to be an increasing demand for roadmapping the future with varying foci and objectives. Roadmapping is considered both as a line of strategic thought and as a process methodology. First, roadmapping as a line of strategic thought aims at combining different modes of knowledge with specific activity layers (Kostoff and Schaller 2001; Phaal, Farrukh, and Probert 2004). In other words, roadmaps are strategic tools for the crystallisation and combination of organisational knowledge that might seem 'unlinkable' with other strategic methods. Second, as a process methodology, roadmapping consists of several modules that can be flexibly linked together. Modularisation allows researchers to form a tailored 'response chain' to answer different kinds of research and development problems in different technology areas (see e.g. Lee and Park 2005). Modularisation also makes space for the combination of different research methods (Ahlqvist et al. 2007a, b). Furthermore, modularisation enables the tailoring of the roadmapping process to suit the needs of the different actors and different tasks in the innovation network. In many occasions, VTT roadmaps can be considered as informative processes with diverse perspectives. However, roadmapping studies at VTT are usually connected to some instrumental procedure that embeds the outcomes of the roadmaps to the strategic management and priority setting in the client organisation. Generally, the roadmapping process in VTT is formed of different modules that combine the external future potentials to the internal strategic activities.

Modular design has also been applied in cross-border foresight processes, such as the *Nordic ICT Foresight* project (Ahlqvist et al. 2007a, b) and the *Nordic Hydrogen Energy Foresight* (Dannemand Andersen et al. 2005). In these contexts the modular design offered possibilities to apply different methods, to engage different stakeholders in different phases, and to develop country-specific approaches to better respond to different national conditions. The successful implementation of such a modular process called for careful consideration of dependability between the phases. Responsive management throughout the process supported adequate changes in stakeholder participation and communication when completing the complex scenario-based process.

Table 2 illustrates the meaning of the foresight management dimensions (expected outputs, future perspectives, methods and stakeholder participation) in the context of a VTT roadmapping process. In particular, the generic key modules of a VTT roadmapping process are related to the choices within the various dimensions. Module 1 '*Drivers and bottlenecks*' refers to societal development trajectories. In the roadmapping process, the outcomes of the first module are at the same time specified to a certain theme, i.e. instrumental, and yet simultaneously they form a kind of general view to the changes in the societal drivers and bottlenecks. The future perspectives of the module are based on a mixture of consensus on the most important trajectories and sensitivity to the potential disruptive phenomena that, if realised, could transform the picture drastically. The module can be managed either as a fixed process through, e.g. questionnaires or interviews, or as an autonomous – and hence a more closed – workshop process. Depending on the management preference, the stakeholder engagement can be more open or closed. The second module '*Markets*' is basically quite similar to the first module. However, the topic of analysis focuses on the market dynamics and business environment of the topic area. Also this module can be completed in a more open or closed fashion, depending on the preferences and the topic area.

Table 2. Examples of generic roadmap modules and their foresight management dimensions.

Roadmapping modules	Outcomes	Future perspectives	Management	Stakeholder engagement
Module 1: Drivers and bottlenecks	*Instrumental* Analysis of the drivers and bottlenecks in a specific topic area *Informative* General understanding of the emerging trajectories	*Consensual* Analysis usually based on the formation of consensus *Diverse* Understanding the potential disruptive phenomena	*Fixed* Questionnaires Thematic interviews Delphi *Autonomous* Workshop dialogue Matrices and tables	*Extensive* Evaluation with large stakeholder participation *Exclusive* Evaluation in a closed workshop for a specific purpose
Module 2: Markets	*Instrumental* Market trajectories in a specific topic area *Informative* General understanding of the market dynamics	*Consensual* Market trends, e.g. on potential consumer preferences *Diverse* Understanding market dynamics in specific fields Counter trends Disruptive phenomena	*Fixed* Questionnaires Thematic interviews Delphi *Autonomous* Workshop dialogue Matrices and tables	*Extensive* Evaluation with large stakeholder participation *Exclusive* Evaluation in a closed workshop for a specific purpose
Module 3: Business concepts	*Instrumental* Visionary business concepts in a specific topic area *Informative* Generic ideas for the construction of business concepts	*Consensual* Requires a consensus of the directions of the business concepts *Diverse* Requires diversifying ideas, understanding of the markets and business environment	*Autonomous* Workshop dialogue Matrices and tables Visioning business concepts requires expert knowledge and multiple iterations	*Exclusive* Visionary business concepts are mostly created for a specific purpose and in a specific topic area Requires mostly exclusive engagement of the key experts in the topic area

(*Continued*)

Table 2. Continued.

Roadmapping modules	Outcomes	Future perspectives	Management	Stakeholder engagement
Module 4: Applications/ services	*Instrumental* Visionary applications or services in a specific topic area	*Consensual* Requires a consensus of the directions of the applications and/or services *Diverse* Requires contextual understanding and diversifying ideas	*Autonomous* Workshop dialogue Matrices and tables Visioning business concepts requires expert knowledge and multiple iterations	*Exclusive* Visionary applications are mostly created for a specific purpose and in a specific topic area Requires mostly exclusive engagement of the key experts in the topic area
Module 5: Enabling technologies	*Instrumental* Technologies that enable applications, services and business concepts in a specific topic area *Informative* General enabling technologies that could be applicable in multiple applications	*Consensual* Consensus on the most important enabling technologies *Diverse* Alternative technological trends Technological disruptions	*Fixed* Questionnaires Thematic interviews Delphi *Autonomous* Workshop dialogue Matrices and tables	*Extensive* Utilisation of large pool of technology experts and citizens *Exclusive* Closed evaluation with technology experts in some specific topic area R&D and non-commercialised technologies

The modules 3 '*Business concepts*' and 4 '*Applications/services*' of Table 2 aim to characterise visionary business concepts and applications that could proactively respond to, or even change the whole landscape of the emerging challenges identified in the modules 1 and 2. Modules 3 and 4 differ from the first or the second module also in a sense that they are more exclusive and autonomous, expert-driven modules. The aims of these modules are topically more contextualised, i.e. dependent on the institutional and organisational setting of the roadmapping process. Therefore, constructing visionary business concepts or ideas for applications and services requires mainly organisational consensus on the general directions of the development, i.e. where to focus and where not to focus. Beyond that the ideation and construction of business concepts, services and applications is an expert-driven and organisationally contextualised process requiring space for open dialogue and more systematic iterations.

Module 5 '*Enabling technologies*' is then again a module that builds on both consensus and diversity. Enabling technologies develop both through linear development paths and through radical shifts that can be identified quite feasibly by combining fixed and autonomous management of the process. In this module it is therefore useful to combine extensive and exclusive participation to get the most feasible results. However, the stakeholder engagement might be dependent on the roadmap topic: if the roadmap is dealing with technologies that are critical for the client and still in the R&D phase, the extensive participation might not come into question.

To sum up, Table 2 illustrates how the design and management dimensions differ in different modules of the VTT roadmapping process. Such a modular approach help address case-specific conditions including the diverse expectations among stakeholders.

4.2. *Strategic management of foresight portfolio*

The developed classification framework may provide some relevant starting points for the construction of a broader picture on the foresight activities that an organisation is engaged with. In particular in contract research organisations, foresight activities are often initiated as a response to the on-going policy and business needs, which may lead to a plethora of foresight activities and a need for understanding what these activities mean together and how the projects really differ from each other. Such analyses should help understanding the present and potential future role of foresight activities within the organisation and in the innovation system at large. Towards this end, the framework was attested as one way to position even strikingly different foresight projects in the common framework to provide a holistic understanding of the foresight activities and the further management of a foresight portfolio.

The constructed portfolio of the VTT horizontal foresight expert team illustrates the clear emphasis on informing rather than providing instrumental support for decision-making. This provides relevant input for the strategic discussion on the role of foresight activities, especially, the positioning of foresight in view of decision-making. Furthermore, the quadrant of the instrumental outcomes with diverse perspectives includes only one roadmap project. Is this a natural consequence of the inherent difficulties in communicating diverse perspectives to decision-makers, and should this be a crucial development area for the VTT foresight team? With regard to the consensual and diverse perspectives and the process management, it appears that consensus-oriented projects have the emphasis towards fixed processes and the diverse perspectives oriented projects, in turn, towards autonomous processes. This leads to the discussion on the chosen methods and process management and to the questioning if such emphases are desirable and if particular improvements in the foresight design and management are needed.

5. Conclusions

This paper has discussed a shift in foresight practice and theory from technology-focused forecasting to the participatory foresight on the interactions of diverse technological systems, product and service applications, market mechanisms and societal drivers, with emphasis on forming the future. The focus has shifted from the production of predictive technology lists and objective fact finding to the elicitation of stakeholder opinions on alternative future pathways. A conceptual framework was introduced for the classification of foresight projects. The application of the framework to examples from VTT foresight portfolio illustrated how the different dimensions of foresight management influence the methodological choices and stakeholder participation.

Further to the management of foresight activities at the portfolio level, the application of the framework seemed to raise relevant inputs to the strategic discussion on the role of foresight activities in the decision-making processes and on the management of foresight projects in view of methodological and process facilitation perspectives. However, the role of foresight activities is strongly related to the advances in the whole innovation system. This lays further questions on the understanding of the context in which the foresight activities are implemented. Because, the framework does not explicitly consist of such perspectives, it is not sufficient for the purposes of the evaluation of foresight activities. Rather, it is appropriate as a tool for providing complementary support for the design and management of foresight projects and the portfolio of projects.

The framework presented in this article has value for both the design and management of individual foresight projects and of a portfolio of foresight activities. For example, building on the application of the framework in the analysis of VTT foresight activities, we identified coherent and modular application of foresight methods, and the responsive engagement of stakeholders, as relevant approaches when balancing with the expectations and objectives related to different dimensions of foresight management. In support of building a broader view of the management of foresight activities, the framework was attested as one way to position even strikingly different foresight projects in the common framework of analysis. The framework was tested as a useful tool in the context of a contract research organisation. The test makes us confident to encourage its further elaboration. The elaboration should include both the further theoretical refinement taking into account other strands of foresight and futures research and the further application in contract research organisations as well as in other policy and organisational contexts.

Note

1. The views expressed are purely those of the authors and may not in any circumstances be regarded as stating an official position of the European Commission.

References

Ahlqvist, T. 2005. From information society to biosociety? On societal waves, developing key technologies, and new professions. *Technological Forecasting and Social Change* 72: 501–519.

Ahlqvist, T., H. Carlsen, J. Iversen, and E. Kristiansen. 2007a. Nordic ICT foresight. Futures of the ICT environment and applications on the Nordic level. VTT Publications 653. Edita Prima Oy, Helsinki.

Ahlqvist, T., H. Carlsen, J. Iversen, and E. Kristiansen. 2007b. Nordic ICT foresight. Futures of the ICT environments and applications on the Nordic level. Summary report. Publication of the Nordic Innovation Centre series.

Ansoff, I. 1975. Managing strategic surprise by response to weak signals. *California Management Review* 17, no. 2: 21–33.

Arthur, W.B. 1989. Competing technologies, increasing returns and lock-in by historical events. *Economic Journal* 99, no. 394: 116–131.

Arthur, W.B. 1994. *Increasing returns and path dependence in the economy*. Ann Arbor: University of Michigan Press.

Barré, R. 2002. Synthesis of technology foresight. In *Strategic policy intelligence: current trends, the state of play and perspectives*, ed. A. Tübke, K. Ducatel, J. Gavigan and P. Moncada, Institute for Prospective Technological Studies (IPTS), Technical Report EUR-20137-EN, Seville.

Coffman, B.S. 1997. Weak signal research, Part III: Sampling, uncertainty and phase shifts in weak signal evolution. MG Taylor Corporation. http://www.mgtaylor.com/mgtaylor/jotm/winter97/wsrsampl.htm

Cruickshank, J., and L. Susskind. 1987. *Breaking the impasse: consensual approaches to resolving public disputes*. New York: Basic Books.

Cuhls, K. 2003. From forecasting to foresight processes – new participative foresight activities in Germany. *Journal of Forecasting* 22, nos. 2–3: 93–111.

Cuhls, K., K. Blind, and H. Grupp. 2002. *Innovations of our future*. Heidelberg: Physica Verlag.

Da Costa, O., P. Warnke, C. Cagnin, and F. Scapolo. 2008. The impact of foresight on policy-making: insights from the FORLEARN mutual learning process. *Technology Analysis & Strategic Management* 20, no. 3: 369–387.

Dannemand Andersen, P., M. Borup, K. Borch, J. Kaivo-oja, A. Eerola, T. Finnbjörnsson, E. Øverland, A. Eriksson, T. Malmér, and B. Mölleryd. 2007. Foresight in Nordic innovation systems. Summary report of the Nordic Foresight Forum funded by the Nordic Innovation Centre in 2005–2007. http://www.nordicinnovation.net/_img/nordic_foresight_forum_final_report.pdf

Dannemand Andersen, P., B. Holst Jørgensen, A. Eerola, T. Koljonen, T. Loikkanen, and E.A. Eriksson. 2005. Building the Nordic research and innovation area in hydrogen. Summary Report. Nordic Innovation Centre and Nordic Energy Research. http://www.h2foresight.info/

David, P.A. 1985. Clio and the economics of QWERTY. *The American Economic Review* 75, no. 2: 332–337.

Eerola, A., and B. Holst-Jørgensen. 2002. Technology foresight in the Nordic countries. A report to the Nordic Industrial Fund, Center for Innovation and Commercial Development. Risoe-R-1362(EN).

Eerola, A., and B.H. Jørgensen. 2008. Foresight in the Nordic countries. In *The handbook of technology foresight – concepts and practice*, ed. L. Georghiou, J. Cassingena Harper, M. Keenan, I. Miles and R. Popper. Cheltenham: Edward Elgar.

Eerola, A., and E. Väyrynen. 2002. Developing TF and TA practices on the basis of European experience. VTT Tiedotteita – Research Notes 2174. In Finnish, abstract in English. http://www.inf.vtt.fi/pdf/tiedotteet/2002/T2174.pdf

Eerola, A., T. Loikkanen, B.H. Jørgensen, P. Dannemand Andersen, and E.A. Eriksson. 2004. Nordic H_2 Energy Foresight – Complementary Contribution of Expert Views and Formal Analyses. A paper presented at the EU–US Scientific Seminar 'New Technology Foresight, Forecasting and Assessment Methods', Seville, 13–14 May 2004. http://www.h2foresight.info/Publications/Publications_Conference.htm

Eerola, A., T. Loikkanen, T. Koljonen, B.H. Jørgensen, P. Dannemand Andersen, and E.A. Eriksson. 2005. Nordic H_2 energy foresight – managerial challengers of the interactive process. A paper presented at the 17th Annual International Conference 'Foresight Management in Corporations and Public Organisations – New Visions for Sustainability', Helsinki, 9–10 June 2005. http://www.tukkk.fi/tutu/conference2005/default.htm

Eriksson, E.A., and M. Weber. 2006. Adaptive foresight: navigating the complex landscape of policy strategies. A paper presented at the Second International Seville Seminar on Future-Oriented Technology Analysis, 28–29 September 2006. http://forera.jrc.es/documents/papers/FTA_EAE_MW%20Final%20060807.pdf

European Commission. 2002. Thinking, debating and shaping the future: foresight for Europe. Final Report of the High Level Expert Group for the European Commission, 24 April 2002, European Commission, Brussels.

Grabher, G., and D. Stark. 1997. Organizing diversity: evolutionary theory, network analysis and postsocialism. *Regional Studies* 31, no. 4: 411–423.

Harris, D., and S. Zeisler. 2002. Weak signals: detecting the next big thing. *The Futurist* 36, 21–28.

Helmer, O. 1983. *Looking forward: a guide to futures research*. Thousand Oaks, CA: Sage.

Hjelt, M., P. Luoma, E. van de Linde, A. Ligvoet, J. Vader, and J. Kahan. 2001. Experiences with national technology foresight studies, Sitra Report Series 4/2001, Helsinki, July 2001.

Irvine, J., and B.R. Martin. 1984. *Foresight in science, picking the winners*. London: Dover.

Jacobsson, S., and A. Johnson. 2000. The diffusion of renewable energy technology: an analytical framework and key issues for research. *Energy Policy* 28, no. 9: 625–640.

Keenan, T. 2003. Identifying emerging generic technologies at the national level: the UK experience. *Journal of Forecasting* 22, no. 2–3: 129–160.

Kemp, R., and Loorbach, D. 2006. Transition management: a reflexive governance approach. In *Reflexive governance for sustainable development*, ed. Jan-Peter Voss, Dierk Bauknecht and René Kemp, 103–130. Cheltenham: Edward Elgar.

Kivisaari, S., R. Lovio, and E. Väyrynen. 2004. Managing experiments for transition: examples of societal embedding in energy and health care sectors. In *System innovation and the transition to sustainability: theory, evidence and policy*, ed. Boelie Elzen, Frank W. Geels and Ken Green. Cheltenham: Edward Elgar.

Koivisto, R.A., M.O.E. Reunanen, and M.H. Nissilä. 1997. Safety reports in Finnish process industries: experiences based on some case studies. *Journal of Loss Prevention in the Process Industries* 10, no. 2: 77–81.

Koivisto, R.A., K.M. Törmäkangas, and V.S. Kauppinen. 2002. Hazard identification and risk assessment procedure for genetically modified plants in the field – GMHAZID. *Environmental Science and Pollution Research* 9, no. 2: 110–116.

Kostoff, R.N., and R.R. Schaller. 2001. Science and technology roadmaps. *IEEE Transactions on Engineering Management* 48, no. 2: 132–143.

Könnölä, T. 2006. Escaping path dependence – essays on foresight and environmental management. Dissertation for the degree of Doctor of Science in Technology, Helsinki University of Technology, ISBN ISBN 951-22-8207-0.

Könnölä, T., V. Brummer, and A. Salo. 2007. Diversity in foresight – insights from the fostering of innovation ideas. *Technological Forecasting and Social Change* 74: 608–626.

Könnölä, T., A. Salo, and V. Brummer. (forthcoming) Foresight for European coordination: developing national priorities for the forest based sector technology platform. *International Journal of Technology Management*, (special issue on technology foresight).

Könnölä, T., G. Unruh, and J. Carrillo-Hermosilla. 2007. Toward prospective voluntary agreements: reflections from a hydrogen foresight project. *Cleaner Production. Journal of Cleaner Production* 15: 259–265.

Lee, S., and Y. Park. 2005. Customization of technology roadmaps according to roadmapping purposes: overall process and detailed modules. *Technological Forecasting and Social Change* 72: 567–583.

Linstone, H.A. 2007. The rise and fall of corporate R&D: out of the dusty labs. *The Economist*, 3–9 March 2007: 74–76.

Lundvall, B.-Å., ed. 1992. *National systems of innovation: towards a theory of innovation and interactive learning*. London: Pinter.

Martin, B.R., and R. Johnston. 1999. Technology foresight for wiring up the national innovation system. Experiences in Britain, Austria, and New Zealand. *Technological Forecasting and Social Change* 60: 37–54.

Mendonça, S., M. Pina e Cunha, J. Kaivo-oja, and F. Ruff. 2004. Wild cards, weak signals and organisational improvisation. *Futures* 36, no. 2: 201–218.

Nonaka, I. 1994. A dynamic theory of organisational knowledge creation. *Organisation Science* 5, no. 1: 14–37.

Nonaka, I., and H. Takeuchi. 1995. *The knowledge-creating company*. New York: Oxford University Press.

Phaal, R., C.J.P. Farrukh, and D.R. Probert. 2004. Technology roadmapping – a planning framework for revolution and revolution. *Technological Forecasting and Social Change* 71: 5–26.

Salmenkaita, J.-P., and A. Salo. 2004. Emergent foresight processes: industrial activities in wireless communications. *Technological Forecasting and Social Change* 71, no. 9: 897–912.
Salo, A., T. Gustafsson, and R. Ramanathan. 2003. Multicriteria methods for technology foresight. *Journal of Forecasting* 22, no. 2–3: 235–255.
Salo, A., T. Könnölä, and M. Hjelt. 2004. Responsiveness in foresight management: reflections from the Finnish food and drink industry. *International Journal of Foresight and Innovation Policy* 1, no. 1–2: 70–88.
Smith, K. 2000. Innovation as a systemic phenomenon: rethinking the role of policy. *Enterprise & Innovation Management Studies* 1: no. 9: 73–102.
Smits, R., and S. Kuhlmann. 2004. The rise of systemic instruments in innovation policy. *International Journal of Foresight and Innovation Policy* 1, no. 1: 4–32.
Tushman, M.L., and C.A. O'Reilly. 1997. *Winning through innovation*. Boston, MA: Harvard Business School Press.
Unruh, G.C. 2000. Understanding carbon lock-in. *Energy Policy* 28, no. 12: 817–830.
Van der Meulen, B., J. de Wilt, and H. Rutten. 2003. Developing futures for agriculture in the Netherlands: a systematic exploration of the strategic value of foresight. *Journal of Forecasting* 22, no. 2–3: 219–233.
Voss, J-P., and R. Kemp. 2006. Sustainability and reflexive governance: introduction. In *Reflexive governance for sustainable development*, ed. J-P. Voss, D. Bauknecht and R. Kemp, 3–27. Cheltenham: Edward Elgar.

e-Government 2015: exploring m-government scenarios, between ICT-driven experiments and citizen-centric implications

Gianluca C. Misuraca

College of Management of Technology, Swiss Federal Institute of Technology in Lausanne (EPFL), Lausanne, Switzerland

After introducing the challenges and failures of first generation e-government (identified as e-Gov 1.0), and the rhetoric of the next generation of e-government efforts (identified as e-Gov 2.0), this paper presents an analysis of the promises of m-government, a relatively new phenomenon whose potential is largely unknown and unexplored but is already considered by many to be the most important subset of future e-government services. Discussing some cases of m-government technologies and applications worldwide, the paper identifies a number of risks and research questions to be further elaborated. In practice, are we about to go through a major adapting societal move, partly due to information and communication technologies (ICTs) but also to other co-factors likely to influence our near future (socio-political-economic and environmental trends), or are we just passing through a natural evolution (eventually faster than others occurred in the past)? Within this particular dynamic, the paper further addresses the importance of looking at what the various emerging scenarios may be, and which ones have to be considered and eventually deployed in terms of policy-relevant and favourable cost–benefit perspectives.

1. Introduction: a preliminary attempt to conceptualise m-government

Contrary to what was stated by the World Bank in 2002 ('there is no e-government textbook and no e-government theory …'), e-government has been studied a lot in the early 2000s and from a number of different perspectives; many even argue that it should represent a discipline in itself, notwithstanding its clear multidisciplinary character.

So far, as described by Cordella (2007), 'the dominant literature has seen e-government as a next step in the rationalisation of government activities along the line of the new public management' (from Bellamy and Taylor (1998) to Fountain (2002) and Heeks (2002)). The concept and practice of e-government has also been analysed deeply from an Information Systems Theory point of view (e.g. Ciborra 1993, 2005; Batini, Viscusi, and Cherubini 2006, 2008). More often, e-government is conceived as a non-scientific domain, to be analysed more simply from a practitioner's point of view, and with specific consideration given to the different 'pillars' of the New Public Management (NPM) agenda: efficiency, accountability, decentralisation and marketisation (or client-focused service delivery, where the optimisation of procedures is key in redefining governance practices and processes).

Critical positions opposing this assumption have emerged in the literature (for instance Di Maio 2002; Dunleavy et al. 2006; Rossel and Finger 2008; Misuraca and Rossel 2007; Misuraca 2007; Avegrou et al. 2005; Cordella 2007), pointing out the inherent 'multi-dimensional components and levels of governance that are needed to be taken into consideration when analysing e-government, especially considering it alongside the transformation of the state and dynamic tensions between rapid technological developments and regulations, the institutional frameworks and capacities in a globalised and multi-stakeholder world (Misuraca 2007).

In our approach, e-government (which has been defined in many ways as e-Gov, digital government, online government or, in certain contexts transformational government) represents a middle-of-the road concept, and identifies the composite trend of governments at all levels, mainly through their operational arm, the administration and – subsidiarily – through the access of citizens to public affairs, aimed at promoting: (1) better and more efficient administration; (2) more effective inter-administration and administration–enterprise relationships; and (3) user-empowering servicing and more transparent access of citizens to political decision-making (see Misuraca 2007; Misuraca, Rossel, and Finger 2006).

The conceptual framework we propose with regard to the introduction of information and communication technologies (ICTs) in government, in contrast to the majority of scholars, is based on neo-institutionalism and social constructivism. The focus of the analysis of e-government and m-government is on the way the different stakeholders interact when introducing ICTs and the way these interactions can affect institutions and society at large. At the same time, we also consider how social phenomena develop in the particular social contexts characterised by ICTs and mobile technologies, also assuming that interactions among various stakeholders are done with the understanding that their respective perceptions of reality are related and, as they act upon this understanding, their common knowledge of reality becomes reinforced.

With regard to mobile government (or simply m-government), it is an even more recent concept, and therefore there is not yet a real theoretical discussion in relation to this phenomenon. It can be defined as 'a strategy and its implementation involving the utilisation of all kinds of wireless and mobile ICTs, services, applications and devices for improving benefits to the parties involved in e-government' (Arazyan 2002). It is mainly referred to as 'the extension of e-government to mobile platforms', as well as the 'strategic use of government services and applications which are only possible using mobile technologies (mobile telephone, Personal Digital Assistant (PDA), laptop and wireless internet infrastructure)' (Antovsky and Gusev 2003). One of the main theorists and proponents of m-government (Ibrahim Kushchu) argues that 'As e-business evolves towards m-business, e-government seems to follow the trend with a few but significant mobile government applications' (see Kushchu and Kushchu 2003).

According to many 'm-government optimists', in fact, the main developments of mobile technologies and applications in convergence with Web 2.0 experiments and technologies could in fact represent a disruptive change in transforming relationships among different stakeholders, and redesign institutional settings, thus creating room for the paradigmatic shift that e-government was supposed to bring about.

However, this is far from being a reality, and it is not clear if and how this will be realised. The experiments and the claim that m-government will be the channel for next generation e-government in the years to come, are still limited by the absence of clear results about the uptake of m-government services and their benefits, as well as any quantification of cost savings that can be realised by performing transactions using mobile technologies vs traditional services or e-government services.

At the same time, the debate on transaction costs and the transforming impact of mobile and wireless technologies is currently open and controversial. In addition to the simply substitutive role of mobile technologies vs traditional ICTs, emphasis should be put on identifying which are the new services that have no equivalent in the 'pre-mobile era' and that can fit into the larger 'mobile ecosystem', differentiated by the 'on the go' characteristic that is only possible to have through the use of wireless ICTs. (On this debate see for instance Hyman and Katz 2001; Hayes and Lemstra 2008.)

Therefore, if m-government *per se* cannot be considered as a paradigmatic shift, what we should look at is rather if, within the broader phenomenon of Mobile Internet services development, it will represent an element that could bring about a real change in redefining organisational boundaries, value chain configurations and lifestyles. While indeed it is perceived so by many, we question whether it would be just a new fantasy, a gradual transformation which some actors try to force into the future and with themselves as lead players, thanks to the classical push-type of rhetoric buzz words. Moreover, if a real paradigmatic shift is indeed taking place, to what extent and with what impact? If this is a real change, what would be its supportive indications?

In practice, the main research question we wish to pose here is 'Are we about to go through a major adapting societal move, partly due to ICTs but also to other co-factors likely to influence our near future (socio-political-economic and environmental trends), or are we just passing through a natural evolution (eventually faster than others occurred in the past)?'

2. Methodological approach

Indeed, mobile services and technologies have rapidly evolved. However, the use of these technologies is still in its infancy and remains to be further researched and documented.

Given that m-government is an emerging phenomenon, there is still little research in this area, and it is not clear which theory better explains its implications (mobility issues, transaction costs, technology assessment of emerging technologies, new public management, diffusionism or social shaping of technologies and constructivism).[1]

The methodological approach we propose is therefore a combination of research methods. In this paper we undertook an exploratory research which involved an open-ended study, not restricted by a specific theory and intended to provide a new body of empirical knowledge from which (in a second stage) theory might be able to be enriched. In future research, the objective is to validate assumptions with 'triangulation' of distinct methods for cross-analysis on comparable data (using quantitative and qualitative analysis techniques).

3. The failure of e-Gov 1.0

With the advent of ICTs and their integration in government operation, the e-government phenomenon took place and was seen as the 'silver bullet' that would change the way government operates and transform the relationship with businesses and citizens (on this see Misuraca 2007; and Misuraca, Rossel, and Finger 2006).

Many policy makers, practitioners and researchers claimed for a paradigmatic shift of e-government solutions, as a driver of transformation of the state, thus an almost linear (and automatic) ICT-driven process leading to better services, lower costs and beneficial to all.

G2G, G2B and G2C were considered as the components of the magic equation that lead companies and governments to invest enormously in ICTs, in many cases without even having the possibility of monitoring the return on investment[2] or the socio-economic impact of ICTs and e-government in the 'real world'.[3]

As a matter of fact, behind the high-tech glamour of most complex ICT projects and e-government initiatives lies a dirty reality: the majority of these projects are failures.[4] The introduction of ICTs in government is not producing the expected results foreseen by the 'messianic' announcements made from Al Gore to Bangemann, and this is true in both the industrialised and developing countries. The so-called 'leapfrogging' effects to be realised by introducing ICTs, predicted for emerging and developing countries, is not actually taking place.

Of course, due to the political implications of the cost of such failures, there is very little data about rates of success and actual failure of e-government projects. However, some baseline estimates indicate that about 70–80% of e-government projects are either total or partial failures. This is due principally to the lack of 'e-readiness' and the oversize gaps between project design and on-the-ground reality (known as 'design–reality gaps'), meaning the lack of assessment of needs prior to the implementation of a project. These failures come at a high price for the world's poorer countries and a key problem among e-government practitioners and policy-makers is a lack of awareness of these costs: most costs are intangible; few are ever measured in the event of e-government failure; e-government failures are often hushed up (Heeks 2003; Collins 2007).

This may explain why, despite the high prevalence of failure and its costs, many officials and politicians are still very keen on e-government. This may also be explained by the fact that, after a post-Internet bubble indirect subsidising of the information technology (IT) private sector, especially in the USA and Europe, the public sector and the politicians have now realised that there is no room for further waste of resources.

It is clear, in fact, that ICT development is evolving rapidly and the traditional state is not capable anymore of coping with the fast technological changes introduced by ICT and its multidimensional impact on society, which in theory enabled a new governance arena, reinforcing the effects of the globalisation of first generation (according to Friedman (2006) we have in fact now entered the third era of globalisation).

4. The rhetoric of e-Gov 2.0

Since the high cost and the complexity of e-government and ICT-related activities in the public sector was recognised, governments worldwide have looked for ways to reduce risks and understand the external and internal barriers to e-government,[5] in order to overcome failures and successfully implement changes.[6]

New catchwords and immediate solutions emerged, with Business Process Reengineering (BPR) linked to strategic business and IT alignment being seen as crucial for enabling change, albeit in only the 'first order of change' sphere. To achieve a real 'second order change' more importance had to be placed on change management techniques, focusing on organisational culture and human resources empowerment (the concept of empowerment itself is criticised as a rhetorical and simplistic notion from the 'developed' world and not a real solution to problems of lack of capacity and resources).[7]

Invoking a sort of 'return to the future', many scholars started exploring the possibility of e-governance[8] to realise the paradigmatic shift that e-government did not realise, but just changing terminology, although introducing new concepts and multidisciplinary perspectives even based on a solid theoretical framework, does not necessarily produce practical solutions.

New 'value drivers' have been given significance, not limiting ICTs and e-government to reach efficiency gains but focusing more on effectiveness and 'openness', as well as looking at the socio-economic implications of interventions and the involvement of all stakeholders as a key for success.

In this context, and linking to the explosion of Web 2.0 technologies[9] and innovative researches and experiments about the potential of online community of users (which gave birth to neologisms such as crowdsourcing,[10] or new techniques, such as podcasting,[11] for example), several governments around the world, Canada *in primis* and Germany in particular leading the European Union towards that direction, introduced the concept of next generation e-government, or e-Government 2.0.[12]

In this perspective, the citizen is not only 'the King', as he/she had been indicated during the new public management approach, transferred from the private sector metaphor of client/customer vs citizens/user (that created many discussions in literature and practice); the citizen has now become a co-producer of knowledge and information. Imitating (or trying to imitate) some very expensive and not clear if truly successful experiments, from Nespresso to Wikipedia, governments are attempting to get the citizens/users as part of the 'value chain' and the administrative-decision-making process, either through the creation of online community of users to reach the remote and disadvantaged communities, or introducing fancy and advanced technological platforms on their e-government websites, as in the case of Berlin, where a web 2.0 platform for games has been introduced in the City portal (Berlin2.0).[13]

However, someone could argue whether it is the role of the state to make citizens play. Are we somehow coming back to the '*panem et circenses*' minimal role of the State?

5. The promises of m-government (an 'e-Gov 3.0' type of paradigmatic shift or just another fancy catchword?)

Mobile government (m-government) is a relatively new phenomenon whose potential is largely unknown and unexplored. However, it is already considered to be the most important subset of future e-government services. One of the key recommendations of e-GovWorld 2006 organised by a number of International Organizations in South East Asia, was that joint efforts should be done to investigate the use of mobile and wireless technologies for future electronic public services as well as to enhance the reach of existing e-government applications to larger sections of the society. It was also felt that in order to understand user perspectives there is a strong need of dialogue among telecom service providers, government agencies, industry and citizens as stakeholders.[14]

A number of actions have been started in this field, including, for instance, the creation of a knowledge portal and observatory on m-government to follow the developments in the field,[15] as well as starting a discussion group, titled mGov, to enhance networking and knowledge sharing between practitioners, academia and researchers from all over the world.[16] These are just examples of the interest that is surrounding this concept.

Indeed, mobile services and technologies have rapidly, and in some cases astonishingly, evolved. Let us look for example at the exponential deployment of mobile telephony in some European countries where the uptake of mobile telephony has been quite striking, and also in emerging countries in North Africa and Asia.

Table 1. Illustrative country comparison of mobile telephony rapid growth and uptake.

	Fixed line teledensity		Mobile telephony density	Global density
Country	2000	2005	2005	2005
Algeria	5.79	7.82	45.24	53.06
Morocco	4.96	4.26	39.37	43.63
Tunisia	9.99	12.47	56.32	68.79
Egypt	8.64	14.04	18.41	32.45
South Africa	10.88	9.97	19.36	29.33
Nigeria	0.49	0.93	14.13	15.06
France	–	59.01	79.44	138.45
Spain	–	42.91	96.81	139.72
Sweden	–	86.71	93.31	180.02
Italy	–	41.98	124.28	166.26

Source: ITU. 2006.

As an example, Table 1 shows a comparison of some emerging and European countries (see the case of Italy for instance, where each citizen holds at least one mobile phone).

At the same time, several experiments have been initiated worldwide to make best use of mobile and wireless technologies towards better service delivery, including eventual co-production of solutions. For example, with the launch of 3rd generation technologies (and the future envisaged 3.5 and 4th generation of mobile technologies), mobile devices are going to be equipped with additional features of larger computing power for the taking care of voice, text and multimedia content. This provides a potential opportunity for government agencies to explore the ways to enhance the outreach of e-government services with the use of mobile and wireless technologies.

This is the case, for example, of Dubai, where the government has launched an initiative called m-Dubai, to provide information and services to the citizens on mobile phones,[17] or of Malta, where a specific m-government strategy has been designed.[18] In Luxembourg the City Government has launched the 'eCity' initiative, deploying mobile payment for governmental services and aiming at creating a virtual city based on a sort of 'Wireless Internet Area Network' for the overall city.[19]

This type of initiative is rapidly growing and it is most likely only a beginning. As a result of the deployment of an overall trend in the building of the information society, bottom-up and user-driven initiatives are going to spread in an increasingly pervasive manner (like the evoked 'Web 2.0' trend, for instance), playing with light practices, multiplicity of channels, real-time reputation building and informal benchmarking, to force more culturally congruent private–public interaction forms to survive rather than others. This ongoing process is also likely to benefit from upcoming technological convergences, such as the one concerning seamless interoperability between the wireless domain and that of the mobile telephony (the 'next generation' leitmotiv, also known as Mobile Internet).

In this regard, however, in addition to acknowledging the converging processes of various kinds, we should also observe the existence of diverging dynamics (both positive and negative), constantly making futures for alternate options (high performance device diversification, combinations that convey new potential, community usages and new business models, lifestyle that evolves at their own pace, social divides that grow, risks that do not disappear, etc.).

Moreover, while mobile government is seen by many as the next paradigm in delivery of public services, many countries are still lagging on the vision to implement these services which have the far reaching goal of enhancing the service-delivery, for example, to the rural areas and disadvantaged people, the main 'victims' of the digital divide.

The mobile phone has in fact become a symbol of ubiquitous access and eventual prosperity in many emerging and developing countries; but what is its true value to their economies?

According to a study on the wireless markets in China, India and the Philippines (Enriquez, Schmitgen, and Sun 2007), the economic impact of all wireless activity in these countries is up to four times the value to the wireless operators alone. An estimate of the economic benefits of wireless activity in fact, must include not only wireless operators but also auxiliary players and end users. Much of this value appears to come from the productivity gains and economic surplus that wireless customers receive simply by using their mobile phones. These results, according to many, should therefore encourage regulators and operators to get mobile phones into more hands more quickly.

Currently about 1 billion users worldwide are actually part of what is defined (sometimes in a very simplistic manner) the Information Society, mostly connected to the Internet in OECD countries, but how does one realistically reach the next billion users and where will they reside? How can one really bridge the digital divide?

In this regard, m-government should not be seen just as a specific area of e-government (limited to the notion of mobile access), but on the contrary take upon the current dominant movement in favour of mobile technology usages, and steer experiments and initiatives in a way that ultimately benefits, and even better empowers the users and citizens in their various flexibility needs. Mobile Internet is seen by many, in fact, as a potentially powerful change factor, not only in terms of technologies, but especially in the way it can modify the relationships among actors with a more citizen-centric focus.

6. Are we in the presence of a new changing factor in the e-government world?

To the average mobile-phone user, the benefits of mobile communication seem to be obvious. Considering the numbers, worldwide, about one billion people purchased new handsets in 2006. In Algeria, for example, 21 million mobile phone subscribers out of the 32 million population in less than 5 years demonstrates the exponential growth of mobile telephony in emerging economies.[20] In Afghanistan, according to *The Economist*, the wireless population has grown from 20,000 users 5 years ago to 1.3 million today. In China, many of the nearly 100 million rural migrant workers rely on a mobile phone not only for their work, but to stay connected with the families they have left behind. The explosive growth of wireless subscribers – 1.2 million new users added daily in emerging economies around the globe – is leveraging new economic options for society. Generally speaking, users are captivated by the freedom, security and enhanced productivity that mobile phones make possible (McKinsey & Co. 2006). However, the use of wireless technologies is still in its infancy and remains to be further explored and documented. Attaching figures to the value of these benefits is difficult, as it is identifying exactly how society gains from all those handsets in pockets and purses. On balance, all surveys emphasise a productivity increase, considering that a dynamic wireless industry seems to be a benefit to any economy, but for the moment, there is little hard data available to quantify its current impact on productivity, and even less information on the public policies and industry actions needed for its potential value to be achieved. Moreover, beyond the mere substitution rationale, there are not yet appropriate business models to be deployed, perhaps because of the lack of multi-disciplinary approaches.

An overview of the mobile penetration in relation to income of countries worldwide is reported in Figure 1.

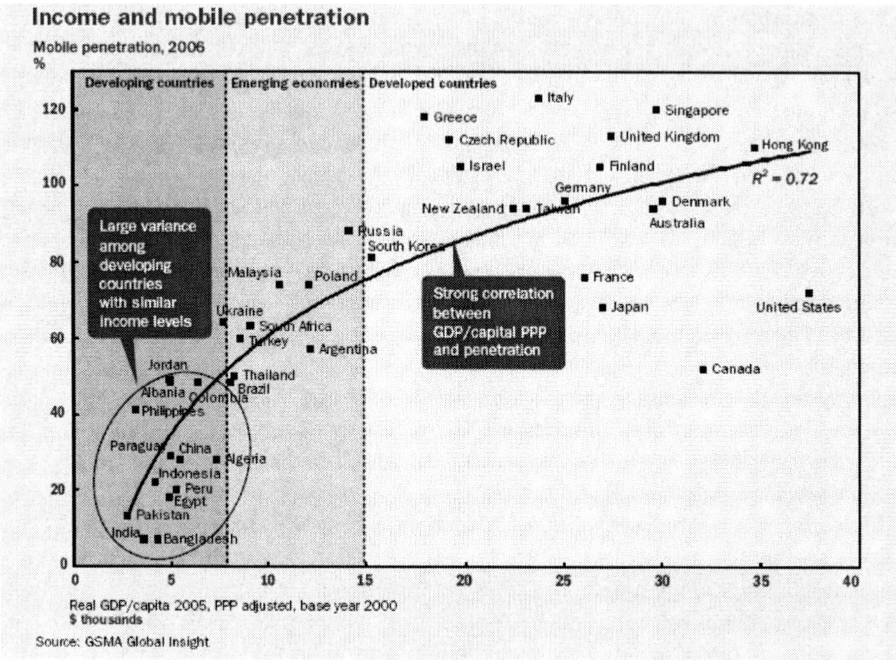

Figure 1. Income and mobile penetration in 2006 (source: GSM Association, see http://www.gsmworld.com/index.shtml).

What really lies ahead of the current achievements in terms of mobile penetration is not so much the ubiquitous access paradigm, which somehow constitutes a base line for more ambitious goals (or perhaps new nightmares if we think of the excesses of use of mobile telephony, its 'intimacy' and the role these devices play in our daily-life, sometime even substituting real relationships), than a new cultural horizon for which administrations and governments will have to adapt, much beyond the e-government evolutionary stage model.

As a matter of fact, some scholars suggest a refined model to understand what is taking place, involving vertical interactions and forms of integrations and overall a pivotal role of m-government in the deployment of e-government (m-government as a tool to obtain higher forms of efficiency, effectiveness, openness and proximity).

Some interesting experiments have been initiated exactly in this direction. For example, New York City is planning to allow mobile phone users to send text and multimedia messages to 911 and 311, and a project in Washington DC, which publishes District-provided road work data overlaid on a Google map, represent two cases of the changes in the way citizens can interact with government using mobile devices. Displays showing the status of pothole repairs effectively keep the city's Department of Public Works on its toes. These examples also illustrate an important aspect of the Web 2.0 world. Some call it 'sousveillance', which happens when people turn the tables and monitor government activity.

Another area where m-government is taking place is the integration of mobile technologies and applications (such as the global system for mobile communications (GSM), for example) combining with geographical information systems (GIS) and global positioning systems (GPS). This is already a reality in some countries that are experimenting with these technologies, in particular at the local and municipal level, in order to provide real-time information about local transportation, directions with maps integrated in the mobile devices, traffic monitoring and crisis prevention, or even indications about the locations of cameras to monitor traffic speed or others (see the cases of Barcelona, Tampere or Switzerland for example). Another phenomenon that is growing is related to the integration of cultural information into mobile devices (see the experiment in the city of Barcelona again, about multimedia information regarding museums integrated into a PDA); somehow reproducing with new technological combinations something that had been proposed as early as the 1990s with iMode in Japan.

Somehow mobile devices have the potential to change people's interactions with government according to the growing scientific understanding of the principle of 'swarm intelligence' that is an important aspect of this potential transformation of government. The term suggests that groups of people may be capable of a higher level of collaborative behaviour than could be predicted from the abilities of individual members.

However, serious concerns about the potential of m-government rely on the limited content so far developed in this area, and the still weak capacity of mobile devices to really act as a substitute to traditional e-government channels, such as computer screens. In this regard, a fundamental component of success will be the way in which the private sector will be associated to the public sector to develop m-government services, in both content design and delivery process. Improving media-streaming and broadcasting functionalities will be pivotal, as well as the large use of m-payment systems, in order to expand the number of usages and users, that are already rapidly increasing as a result of online games and TV-on-demand, two of the central services so far used and soon to be made available on mobile phones, following the success of 'mobile music' with iPod and iTune (see the rapid deployment of iPhone).

At the same time, governments should develop their own strategies and policies to anticipate potential change that will inevitably come because of user pressure and increased involvement of citizens, facilitated in turn by the development and channelling of content in real time provided directly by the users themselves (let us think about the U-Tube approach for example).

The mix of mobile and web 2.0 technologies into a Mobile Internet paradigm, is potentially leading to new forms of participation pushed by enhancement of social awareness, as well as capacity of users. This is more and more the case in emerging markets, where the demographic growth is exponential, and education is also increasing, thus making the availability of a large population of young and skilled people ready to use the latest device, and ready to be part of what someone called the 'ubiquitous society' (see for example u-Korea, u-Japan, etc.):[21] a society where mobility of content is required and mobile access is pervasive and considered as a pre-requisite (thus often given for granted).

This emergent behaviour enabled by the convergence of personal communications and publishing technology with massive private and government data sources could empower individuals in their relations with governments, but this remains to be verified by actual observations and facts.

Smart mobile phones, PDAs and other communication devices, all connected to Web 2.0 applications on the Internet, have the potential to utterly transform the way people react to natural disasters, terrorist attacks, or even ordinary fender benders. This was the case, for example when private citizens collaborated during the aftermath of Hurricane Katrina to share information, help reunite families and pick up the slack where government failed; or after the 11 March 2004 terrorist bomb attack in Madrid, when citizens exchanged information regarding the government position before the upcoming election using mobile telephones (and someone argue this made change opinions of vote for many).

The same happens more and more with regards to civic engagement. The 'mobile trends' are highly promising for organising civic action campaigns and engaging citizens on issues that matter most. For example, activists have used mobile phones to swing elections through innovative get-out-the-vote activities, to ensure impartial voting through poll monitoring, to organise collective action to free political prisoners, and to lobby legislators to pass environmental laws.

A concrete example is the case of Ghana where, during the electoral process in 2003, the opposition equipped some observers in each electoral building with mobile telephones to inform in real time the results of votes that were then diffused via local radio, so to avoid eventual 'mistakes' when the ballots would have reached the central Presidency Office to be verified (and this resulted in the opposition winning the elections after a long period of 'stability').

Mobile phones have been used to mobilise hundreds of thousands of peaceful demonstrators in countries across the world with text messages and political ringtones. They are being used in environmental campaigns in Argentina and to advocate for an independent judiciary in India, for support of emigrant and migrant communities in the Philippines and for increased disease detection and prevention in Rwanda.

In short, there is significant innovation and a new tool for use in social change campaigns. A global network (titled MobileActive) of people, tools, projects and resources focuses on the use of mobile phones for activism, campaigns, and civic engagement has been created. This group is actively engaged in expanding access to knowledge, ideas and experience supported by the use of mobile technology; reducing learning costs for civil society organisations; accelerating the use of effective tactics in campaigns; and providing a comprehensive platform for building partnerships, and for facilitating access to technology and funding. Probably a new business model around mobile technology and civic engagement has been defined.

All these may be identified, in our opinion, as weak signals of an expected evolution (a heavy trend), reinforced by the fact that mobile technology can guarantee user proximity, easier accessibility capacity, and real time access, thus have the potential, in theory, to leverage effects of usage and, once Mobile Internet is a reality, it may rapidly increase the access to Internet, so as to reach the next billion Internet users in a shorter period of time.

In practice, although the adoption and the experience of mobile web access are improving at an impressive rate (mainly in the Organisation for Economic Co-operation and Development (OECD) countries such as USA, as well as UK and Italy that are also growing fast),[22] a report from ABI Research (July 2008) shows that mobile subscribers rate Internet access as one of the least desirable features to have on their handsets. The ABI report contends that this result underscores the difficulties that operators all over the world are having in convincing users to buy paid content. In other words, looking at USA only, more than twice the number of those that could use the mobile web chose not to do so. In practice, many mobile data and multimedia services are failing to reach the mainstream not because they are unavailable, but because they fail to provide a satisfactory user experience and pricing model to most consumers.

7. Going global locally – mobility vs usefulness and effectiveness

In all this, however, lies a fundamental paradox linked to the growing positive (or even positivistic) value given to the perception that being mobile and 'ubiquitous' is generally a plus. This is actually not always the case, although the current societal trends, especially in industrialised countries, want 'knowledge-professionals' to be connected 24/7 (24 hours a day, 7 days a week, and eventually 365 days per year), wherever they are.

The current institutional designs are not adequate. Let us imagine working full time out of the office. What will happen when your boss (or often bosses) ask(s) you where you are? Answers will now be 'I'm online', no matter where I am. A certain degree of face-to-face meeting is still required. An additional problem is the technical limitations of storage on remote servers, when you are for example travelling for more than a month and receive an average of 200 emails per day.

All this has an influence on the effectiveness and decision making capacity as well. Let us think for example of the case – becoming more and more real – of 'Globetrotter Professors', travelling from one conference to another all over the world, and with only email (or Blackberry) rapid communication with their office.

All this has a negative impact on productivity, both scientific and administrative, as all-important decisions still require the paper signature of the Professor, or they may need a brainstorming session that cannot take place only and always on Skype, for example. This has a negative effect on the organisation that the Professors lead, as they would be unable to coordinate even their small office effectively despite perhaps hiring some 'watchdog' to check if employees are at work (or at work but not doing anything effectively for example – while others may be travelling and be more productive being equipped with mobile technologies and, especially, the capacity to use them efficiently). This can also produce a 'rebound effect', for instead of reducing mobility, ICTs are actually increasing the opportunities and the need of travelling, thus increasing costs, limiting productive time (unless you can work in a second class seat on an intercontinental flight, no matter whether you have jet lag or there are babies crying), and ultimately reducing quality of life (on this, see among others Rossel 2003; Mokhtarian 1998; Scientific American 1997).

Despite the hilarious – but still serious – concerns illustrated by these situations, which are applicable to several environments of course, not just Universities, we should consider that this paradox of the nomadic or ubiquitous society brings a number of risks. It can influence not only the productivity and effectiveness of working organisations, but also the human dimension of relationships, mediated by ICTs, that can produce unexpected effects, sometime creating misunderstandings and possible irreparable dangers as a result of an entropic explosion and a waste of energies for coordination and control caused by a lack of clear communication rules (Misuraca 2006).

On the other side, mobile technologies can (and have in principle the potential to) be more beneficial at the local level than at the global scale. In fact, if we talk about m-government in particular, it is more likely that m-services provided by local administrations are more useful to citizens than some national services that eventually are used only once a year. In this regard, m-services can increase the degree of freedom and the proximity of local administrations with citizens. This can create effective opportunity for local economic development, especially in emerging countries where population and ICT capacity is growing fast. The argument is beginning to be made that this could also automatically lead to an improvement of the quality of the governance systems, by 'democratising' through the 'democratisation' of mobile and wireless technologies, the institutional setting itself. However, even if this could ideally be a goal *per se*, it does not actually seem to be realistic, or even possible without other more profound changes that go beyond the scope of this paper of course. What appears to be realistic is to explore future scenarios with a knowledge-intensive society perspective, as an ultimate objective and opposite to a simple (and risky) information-intensive society. In this regard, it is important to outline the fundamental research question we have identified above, which can be reformulated in broader terms, as follows: what kind of changes in society will take place in the next 10–20 years and what will be the role of ICTs and mobile technologies, in particular, in government operation?

The key trends identified in an interesting research by Howkins and Valantin (1997) (and still relevant despite its date), can be distilled into two basic uncertainties related to the global community and national policies. Each uncertainty raises new questions. With regard to the global community, will its value systems become more inclusive and open, or more exclusive and closed? With regard to individual countries, will they have a complete or partial (proactive or reactive) response to acquisition and use of ICTs?

Basic assumptions here are that the global community has two value systems jostling for power. 'One is inclusive, open, and enabling with solid attempts to integrate the weak and disadvantaged. The other is exclusive, fractured, restrictive, a Darwinian world, red in tooth and claw, dominated by the strong and powerful for their own ends, with increasing concentrations of technology, wealth, and power and little regard for those left behind. The difference is not simply one between private and public. It is more the difference between cooperation and exploitation.'

'In turn, national responses range from being complete and positive to being partial, disengaged, and reactive. Some countries have ambitions to learn and create strong policies at home and take a full part in global debates and negotiations. Some want to respond but lack the resources. Others are unresponsive to the challenge. Again, this need not be a split between government and the private sector. Governments must lead, but the whole country must learn'. Given these two uncertainties, the research devised four scenarios (as indicated in Figure 2).

Each scenario starts with the same certainty that technological innovation will continue for the foreseeable future. The dynamism of the information and communication industries will continue to fuel a roller-coaster ride of research, investment, development, acquisition and wealth creation. These companies' shareholders and managers are focused, aggressive and visionary. These factors, more than any public policy, are the drivers of change, but then each scenario diverges according to the two uncertainties defined above. While the levels of technology expand under all scenarios (a key certainty), the levels of access to technology and its potential benefits vary (a key uncertainty). The interaction between certainties and uncertainties, according to this research, will determine the future.

More than ten years after this scenario-building exercise, the future (now present) is yet to be clarified and actually seems to be rather more complex, especially as a result of the rapid and

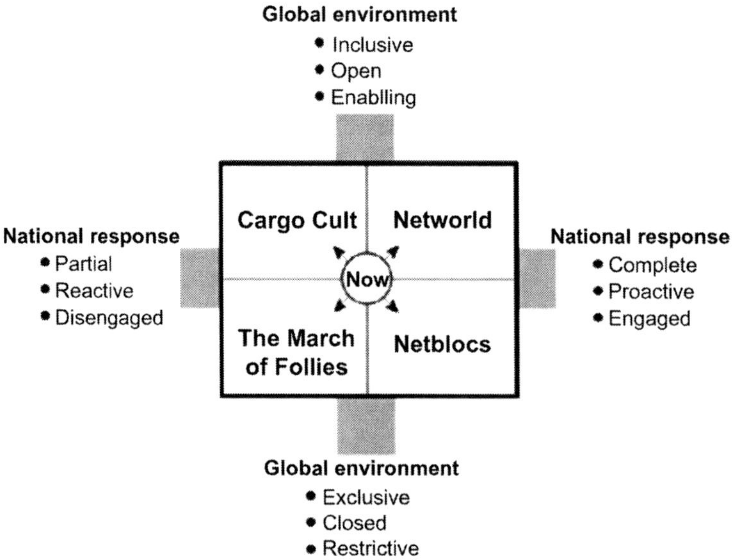

Figure 2. Four global scenarios for the future of ICTs (source: www.idrc.ca).

co-evolutionary development (and deployment) of ICTs and technologies that 10 years ago were not even imaginable.

So, what is likely to be the future in what is now preferred to be called the Knowledge Society? Finally, looking at a crucial issue that could limit and resphape the overall research framework: who will lead the changes at 2015 and further horizons? Will the USA, Europe and OECD countries still be the key players, or will BRICs (Brazil, Russia, India and China) definitely take over (Wilson and Purushothaman 2003)?

8. Conclusions and way forward

Within the particular dynamic we have introduced, it is now important to look at what may be the various scenarios that are emerging, or that are possibly to be considered and eventually deployed in terms of policy-relevant and favourable cost–benefit perspectives.

What about the use of new forms of knowledge development (such as podcasting) for education purposes? What about using mobile devices for disseminating content related to education or other relevant information, instead of using the potential of mobile technology today, and Mobile Internet tomorrow, only for fun?

New research fields are therefore open to explore the kind of changes that m-government is likely to produce; and this at least in five main areas:

(1) In terms of information management, as it will change the way in which information is represented/managed/shared;
(2) In terms of technical challenges that will emerge, for example, simply with regard to new forms of data storage required, or a new way of distributing access within a network, and so on;
(3) In terms of organisational forms, business models (see for example Web 2.0), that may end up in innovative public–private partnerships for instance.
(4) In terms of governmental agents with, for example, the 'efficiency gain' as a result of the 'mobility' of agencies and users, or the expansion of options and forms of use, especially in combination with other technologies or types of applications (mash-ups).
(5) In social terms, as it will increase the risks and the potential threats of information dissemination with regard to privacy and security concerns, for example, as new mechanisms will be required to cope with the new forms of community of users, with consequences also in the way the communities will organise themselves, and eventually new forms of divides that may arise.

Specific research activities, and more and more research networks, are being established and proposed. For example, we at EPFL are among the initiators of several researches in the broad – and not yet well defined – area of e-governance.[23] In this framework, for instance, based on the assumption that e-government has evolved as a field with increasing research and technology development investments over the past 10 years and considering that the use of ICTs in government has become an increasingly important factor for competitiveness and growth in the European Knowledge Society, it is proposed that e-government research should accompany modernisation of the public sector in several ways, and more specifically through: developing methods and techniques to design e-government services; designing proper tools and methods for advancing solutions in the field; assessing the progress and impact of e-government implementations;

and creating the baseline of literacy, knowledge and skills for different stakeholders to use the instruments developed.

As a result of the introduction of e-government and ICTs, in fact, most European Union (EU) member states revised their existing strategies for public sector modernisation and transformation towards reaching innovative ICT applications. These national strategies have been streamlined to meet the common objectives of the EU, such as the i2010 Strategic Framework and the Lisbon agenda, where e-government is recognised as a key initiative for European policy. It has to be noted however that, in most cases, national strategies and action plans are short to mid-term oriented, and the effective implementation is not secured. Therefore, considering the current status of research and maturity of implementation, several deficiencies and imbalances can be identified and must be addressed. Within this context, the role of the private sector, in partnership (and not just as service provider) with the public sector, but also taking into appropriate consideration citizens, as both consumers and co-producer of information and knowledge, will be crucial.

In doing so, governments should focus towards an horizon that is at the same time not too short-term, in order to identify policies and activities that will be able to produce results that can be further replicated and eventually generalised and, also, not too long-term, as technologies (especially in the Mobile Internet arena) evolve very rapidly and may become obsolete before being assessed. Several scenario-building exercises have been initiated, in some cases also using innovative approaches, as blogs or virtual community, starting from the most famous Gartner Research on government 2020.[24]

Of course, despite all the experiences and exercises, and the overview we have tried to provide within this paper, we do not have answers to all the questions raised (although we wish we had). Some of these questions will be answered in the years to come. Others may need to be reformulated and reconsidered according to what decisions will be taken in the coming years with regard to technological choices and new developments, but also and especially with regard to regulatory and policy making decisions, and the uptake of users in innovative experiments, and how again these may shape technological solutions.

At this stage, and as a result of this exploratory research and field-work evidence, m-Services domains can be classified, in our view, in four main categories: (1) activities in which enterprise can get a real gain in flexibility, number of options and easiness of contact with the administration; (2) activities in which mobile workers of enterprises can do their job with increased efficiency; (3) activities that give access to ubiquitous services by consumers/citizens; and (4) activities of those agencies and agents of the administration who have to be mobile to provide efficient servicing (see also Rossel, Misuraca, and Finger 2006, 2007).

Thus, as a preliminary conclusion, m-government appears to be a fuzzy yet promising area for research and experiments in the sense that it opens new perspectives for delocalised governmental servicing and opportunities for more proximity and user involvement in some of the new services offered.

However, several open-ended questions still remain undecided. It may be therefore useful to decompose the problem and look at how to identify the changes that mobile services will bring about. This involves which benefits and indicators will be needed to assess impacts against both positive or negative externalities that can be produced, considering at the same time the value of new services, and not just as an expansion to traditional ICT channels and practices (for example, including face-to-face, but also intermediation mechanisms).

Further research is in particular required, through the specific focus of possible e-government scenarios in order to contribute to outlining appropriate answers. In this inter-linkage most probably lies the substance that will or will not make possible some meaningful and effective transition between the current Information Society trend and more knowledge-demanding forms of societal solutions. In between, there are dynamic tensions to examine (between so-called good practices and innovation inclinations, or between local knowledge and needs and global pressures, and even more immediately, the remapping of the e-government and ICT domains, so as to establish key priorities, eventually helping improve policy planning capabilities and trigger innovative moves in this area.[25]

The time scale of this scenario-building exercise should be up to the year 2015 (but with a longer-term horizon of 2020/2025). This may be the right timeframe considering the strategy-development efforts that many countries, both industrialised and emerging, have currently embarked upon and the technological development path on which mobile operators and developers have been focusing on in the last years. At the same time all main socio-developmental indicators, such as the Millennium Development Goals and the benchmarking indicators agreed by all UN member states during the World Summit on Information Society (WSIS) point to 2015 as a date for assessment of global and national results in achieving a more inclusive and equitable Information Society worldwide.[26]

Notes

1. On technological assessment and diffusionism, see for example Rogers (2003), Dosi (1982), Bass (1986) and Davis (1989); on constructivism and social shaping of technologies, see for example Sorensen (2002), Sorensen and Williams (2002).
2. According to Gartner (2005), in the USA in 2004, the private sector spent US$34 trillion on ICTs, and only a third of the companies made a business case of the expenditure.
3. Some data on the financial expenditure on e-government in the EU and its possible socio-economic impact have been identified by the EU eGEP Project, led by RSO Spa and LUISS Management, http://82.187.13.175/egep/asp/E_Home.asp
4. See for example Heeks (2003), part of an eGovernment for Development Information Exchange project, co-ordinated by the University of Manchester's IDPM, funded and managed by the Commonwealth Telecommunications Organisation as part of the UK Department for International Development's 'Building Digital Opportunities' programme. http://www.egov4dev.org/ or http://www.e-devexchange.org/eGov/topic1.htm
5. On this issue see also: 'Breaking Barriers to eGovernment', an EU-funded project led by the Oxford Internet Institute in collaboration with Gov3 Ltd and the University of Namur, the University of Murcia and the University of Tilburg, http://www.egovbarriers.org/
6. It must be underlined here that importance has been rapidly given to 'ICT use' rather than on its deployment. However, although the use of ICTs accounts for only 25% in organisational management, in many organisations 90% of management efforts are still focused on ICT deployment (see for example Benjamin and Levinson 1993; Benjamin and Blunt 1992).
7. On the concept of Change I to Change II, see for example Rossel, Misuraca, and Finger (2006) and other research activities by Misuraca and Rossel within the framework of the Executive Master in e-governance and at the College of Management of Technology of EPFL and, in particular, Misuraca and Rossel (2007).
8. e-Governance is generally referred to as the capacity of ICTs to harness changes, looking not only at the increasing use of ICTs as a technological tool for delivering services online and improving the efficiency of administrations,

but as a new paradigm for opening up government services to citizens, thereby increasing transparency and participation, and making government more responsive and centred upon its citizens' needs, but this is in my opinion a limited definition of the potential of e-governance. In fact, the conceptualisation of e-governance that I propose is that e-governance concerns a rapidly evolving multi-dimensional, multi-actor, multi-level and inter-sectoral area of application, influenced not only by the ICT-revolution, but mainly by globalisation, international economic competition and state transformation. It also takes place in a very specific environment and contextual atmosphere about which we must understand and learn, in order to best guide the process. This overall change can be identified as manifold, although producing a composed impact on our lifestyle, level of wealth and relationships, inter-regionally and inter-nationally speaking. In short, We define e-governance as 'governance with and of ICTs'. See Misuraca (2007).

9. On the issue of Web 2.0 Technologies, see Glassey (2008, forthcoming). On the topic of Web2.0 and e-government, see also EU-JRC-IPTS (2007), Pascu et al. (2006, 2007).
10. Crowdsourcing is a neologism for a business model in which a company or institution takes a job traditionally performed by a designated agent (usually an employee) and outsources it to an undefined, generally large group of people in the form of an open call over the Internet. The main advantage of crowdsourcing is that innovative ideas can be explored at relatively low cost. Source: Wikipedia (2007).
11. The term 'podcast' is a portmanteau of the name of Apple's portable music player, the iPod and broadcast; a 'pod' refers to a container of some sort, and 'cast' to the idea of broadcasting. In other words, a podcast is a collection of files (usually audio and video) residing at a unique web feed address. People can 'subscribe' to this feed by submitting the feed address to an aggregator (such as iTunes software that runs on the consumer's computer). When new 'episodes' become available in the podcast they will be automatically downloaded to that user's computer. Unlike radio or streaming content on the web, podcasts are not real-time. There is no live broadcast of content, the material is pre-recorded and users can check out the material at their leisure, offline. Source: Wikipedia (June 2007).
12. See http://www.staat-modern.de/E-Government/-,13073/E-Government-2.0.htm
13. See http://www.berlinzweinull.de/index.html
14. See http://www.egovworld.org/
15. See http://www.mgovworld.org
16. See http://in.groups.yahoo.com/group/mGov/
17. The m-Dubai initiative was launched in September 2005 and has put Dubai among one of the governments in Middle East that have taken the lead in m-government initiatives to enable payment-based transactions of e-government services on mobile phones.
18. See http://www.mobile.gov.mt/pl_mitts.asp?mb:lang=en
19. See http://www.vdl.lu/Mairie+et+services+centraux/e_City-style-m.html
20. A specific analysis of the situation in Algeria has been conducted by the author in the elaboration of the 'e-Algèrie 2015' (Misuraca with Abida 2007).
21. See u-Korea IT839 Strategy, http://www.digitaltechnews.com/news/2005/07/ukorea_it839_pr.html; and u-Japan policy, http://www.soumu.go.jp/menu_02/ict/u-japan_en/index.html
22. According to a report by Nielsen (2008), the number of Americans accessing the Internet from mobile handsets has increased from 22.4 million in 2007 to 'some 40 million' in mid 2008.
23. We can mention here files of all the 'Executive Master in e-Governance', an executive training program organised by EPFL in collaboration with prestigious institutions worldwide, and of which this author has been the Managing Director and Scientific Coordinator in the first two editions 2005–2007; as well as other training and research initiatives, including for example the EU Projects eGOVERNET, eGOVRTD2020, etc.
24. http://www.gartner.com
25. This discussion is, in fact, at the basis of a research initiated within the framework of the 'Executive Master in e-Governance' at EPFL and of other researches at the College of Management of Technology at EPFL, and that will be pursued under the auspices of two main research supportive activities, namely, COST Action 22/Swiss funded research on the future of ICTs, and the PhD research of this author. The proposed Research plan includes, building upon the findings from past and current fieldwork analysis in developing and emerging countries, to substantiate and address the above-mentioned research questions. The methodology of the research will be based, on one hand, on research already conducted at EPFL-CDM, and in particular on the various natures of changes taking place under the impact or involving the development of ICTs and, on the other hand, on more specific future studies and technology foresight methodologies, ending up, in particular, in the elaboration of a diversified set of scenarios to be evaluated.
26. On this see: http://www.un.org/milleniumgoals/ and http://www.itu.int/wsis/index.html

References

ABI Research. 2008. Mobile Internet devices and UMPCs: market analysis and forecast. http://www.abiresearch.com/products/market_research/Mobile_Internet_Devices_and_UMPCs

Antovsyki, L., and M. Gusev. 2003. New mobile payments model. Paper presented at the Proceedings of the International Conference on Advances in Infrastructure for Electronic Business, Education, Science, Medicine, and Mobile Technologies on the Internet (SSGRR 2003), July 28 – August 3, L'Aquila, Italy.

Arazyan, H. 2002. m-Government: definition and perspectives. Development Gateway. http://www.developmentgateway.org

Avegrou, C., C. Ciborra, A. Cordella, J. Kallinikos, and M. Smith. 2005. The role of information and communication technology in building trust in governance: towards effectiveness and results. Research Report. Inter-American Development Bank, Washington, DC.

Bass, F.M. 1986. The adoption of a marketing model: comments and observations. In *Innovation diffusion models of new product acceptance*, ed. V. Mahajan and Y. Wind. Cambridge, MA: Ballinger.

Batini, C, G. Viscusi, D. Cherubini, and A. Maurino. 2007. A methodology for the planning of appropriate e-government services, in *Interdisciplinary aspects of information systems studies*, The Italian Association for Information System. ed. A. D'Atri, M. De Marco and N. Casalino. New York: Springer. http://www.springerlink.com/content/q878237093535114/

Batini, C., G. Viscusi, and D. Cherubini. 2008. eG4M: the planning methodology. QD Research Report, University of Milan, Bicocca, DISCO, Starrylink Ed.

Batini, C., G. Viscusi, and D. Cherubini. 2009. GovQual: a quality driven methodology for E-Government project planning. *Government Information Quarterly* 26, no. 1: 106-17.

Batini, C., G. Viscusi, and D. Cherubini. 2008. EG4M. The planning methodology, in Tesi e Ricerca (Report). University of Milano, Bicocca.

Bellamy, C., and J. Taylor. 1998. *Governing in the information age*. Buckingham, UK: Open University Press.

Benjamin, R.I., and E. Levinson. 1993. A framework for managing IT-enabled change. *Sloan Management Review* 34, no. 4: 23–33.

Benjamin, R.I., and J. Blunt. 1992. Critical IT issues: the next ten years. *Sloan Management Review* 33, no. 4: 7–19.

Ciborra, C. 1993. *Teams markets and systems*. Cambridge: Cambridge University Press.

Ciborra, C. 2005. Interpreting e-government and development: efficiency, transparency or governance at a distance? *Information Technology & People* 18, no. 3: 260–79.

Collins, T. 2007. Only a third of government projects succeed, says CIO. *Computer Weekly*, 22 May 2007: 4.

Cordella, A. 2007. e-Government: towards the e-bureacratic form? *Journal of Information Technology* 22: 265–274.

Davis, F.D. 1989. Perceived usefulness, perceived ease of use, and user acceptance of information technology. *MIS Quarterly* 13, no. 3: 319–340.

Di Maio, A. 2002. Towards a wireless public sector. Gartner Research, Bridport, CT.

Dosi G. 1982. Technological paradigms and technological trajectories. *Research Policy* 147–162. http://icc.oxfordjournals.org/cgi/content/abstract/17/3/507

Dunleavy, P., H. Margetts, S. Bastow, and J. Thinkler. 2006. *Digital era governnace*. Oxford: Oxford University Press.

Enriquez, L., S. Schmitgen, and G. Sun. 2007. The true value of mobile phones to developing markets. *McKinsey Quarterly*, February, Web Magazine. http://www.mckinseyquarterly.com/The_true_value_of_mobile_phones_to_developing_markets_1917

EU-JRC-IPTS. 2007. Tutorial on the impact of web 2.0 on eGovernment. Presentation to the 'European eGovernement Conference' Lisbon, 19–21 September 2007, Portuguese Presidency of the EU and The DG Information Society and Media, EC.

Fountain, J.E. 2001. Public sector: early stage of a deep transformation. In *The economic payoff from the international revolution*, ed. R.E. Litan. Washington, DC: Brookings Institution Press.

Friedman, T.L. 2006. *The world is flat: a brief history of the twenty-first century*, rev.exp. edn. New York: Farrar, Strauss and Giroux.

Gartner. 2005. Market focus: IT services opportunities in the government sector, USA, Focus Report.

Hayes, V., and W. Lemstra. 2008. How Wi-Fi is contributing to closing the digital divide gap? *Network Industries Quarterly* 10, no.2. http://newsletter.epfl.ch/mir/index.php?module=epflfiles&func=getFile&fid=94&inline=1

Heeks, R. 2002. Reinventing government in the information age. In *Reinventing government in the information age – international practice in IT-enabled public sector reform*, ed. R. Heeks, 9–21. London: Routledge.

Heeks, R. 2003. Most eGovernment-for-development projects fail: how can risks be reduced? iGovernment Working Paper Series, no. 14, University of Manchester.

Howkins J., and R. Valantin. 1997. (ed.) Development and the information age: four global scenarios for the future of information and communication technology, for the International Development Research Centre of Canada, IDRC and the United Nations Commission on Science and Technology for Development, UNSTD, IDRC.
Hyman, M., and R. Katz. 2001. Alternative industry futures in the global Internet economy. In *Creative destruction: business survival strategies in the global internet economy*, ed. L. McKnight, P. Valeer and R. Katz, 229–239. Cambridge, MA: The MIT Press.
Kushchu, I., and H. Kuscu. 2003. From e-government to m-government: facing the inevitable. Paper presented at the 3rd European Conference on E-Government (ECEG03), July 1–2 at Trinity College, Dublin.
McKinsey & Co. 2006. Wireless unbound: the surprising economic value and untapped potential of the mobile phone, December 2006.
Misuraca, G., P. Rossel, M. Finger, 2006. Governance with and of ICTs: the need for new institutional design in a changing world. *egov: Magazine* 2, no. 5.
Misuraca, G. 2007. *e-Governance in Africa, from theory to action: a handbook on ICTs for local governance.* Lawrenceville, NJ: IDRC/Africa World Press.
Misuraca, G.C. with M. Abida. 2007. e-Algèrie 2015: strategic plan for the development of the information society in Algeria. Paper produced within the framework of the EUROPEAID Project of Technical Assistance for the Development of the Information Society in Algeria, Brussels.
Misuraca, G. 2006. e-Governance and the Kitesurf Approach. In Beyond Tunis: Flightplan 1.5. Global Knowledge Partnership, November 2006. http://www.globalknowledge.org
Misuraca, G., Rossel, P., and Finger, M. 2006. Governance with and of ICTs: the need for new institutional design in a changing world. *egov magazine* 2, no. 5: 36–9.
Misuraca, G., and P. Rossel. 2007. Triggering the governance perspective of e-Government projects: beyond mere digitization of administration services. Background paper for the Third Global Knowledge Conference, Kuala Lumpur, 10–13 December 2007.
Mokhtarian, P.L. 1997. Now that travel can be virtual, will congestion virtually disappear? *Scientific American* X, October.
Nielsen Mobile. 2008. Critical mass: the worldwide state of the Mobile Web. Report. Nielsen Mobile, New York.
Pascu, C., J.C. Burgelman, D. Osimo, M. Ulbrich, and G. Turlea. 2006. The potential disruptive impact of emerging social computing applications. EU-JRC-IPTS European Commission, Joint Research Centre, Institute for Technological Prospective Studies, Seville (Spain), Report.
Pascu, C., D. Osimo, M. Ulbrich, G. Turlea, and J.C. Burgelman. 2007. The potential disruptive impact of Internet 2 based technologies. EU-JRC-IPTS Report. *First Monday*, March.
Rossel, P. 2003. Exploring the rebound-effect of information technologies on mobility systems: clues for a blueprint, Proceedings of the STRC 2003, Monte-Verita, Ticino-Switzerland.
Rogers, E. M. 2003. *Diffusion of innovations*, 5th ed. New York, NY: Free Press.
Rossel, P., and M. Finger. 2007. Regulation and termination rates in mobile telephony: state of affairs and option. Report to Swisscom by PRAXIS Research Europe.
Rossel, P., and G. Misuraca. 2007. Triggering the governance perspective of e-Government projects: beyond mere digitization of administration services; or the Change I to Change II shift issue. CDM-EPFL Working Paper, Lausanne, 2007 forthcoming. http://cdm.epfl.ch/
Rossel, P., M. Finger, and G. Misuraca. 2006. Assessing and steering mobile e-Government options: findings and indication proposals. Proceedings of the 6th European Conference on e-Government, 27-28 April, Marburg, Germany.
Rossel, P., and M. Finger. 2008. Conceptualizing e-Governance. ACM International Conference Proceeding Series; Vol. 232, Proceedings of the 1st International Conference on Theory and Practice of Electronic Governance, December 10–13, in Macao, China.
Scientific American. 1997. The future of transportation, *Scientific American* 277, no. 4 (Special Issue).
Sorensen, C. 2002. Digital nomads and mobile services. *Receiver Magazine* 6, Vodafone Group. http://receiver.vodafone.com
Sorensen, K., and R. Williams, eds. 2002. *Shaping technology, guiding policy. Concepts, spaces and tools*. Cheltenham, UK: Edward Elgar.
Wilson, D., and R. Purushothaman. 2003. Dreaming with BRICs: the path to 2050. Goldman Sachs Global Economics Paper, no. 99, October 2003. http://www2.goldmansachs.com
World Bank, InfoDev and the Centre for Democracy and Technology, 2002. The e-government handbook for developing countries. http://www.infodev.org/en/publication.lb.html

Universities and the emerging new players: building futures for higher education

Attila Havas

Institute of Economics, Hungarian Academy of Sciences, Budapest, Hungary

Universities are long-established organisations, and although they have reinvented themselves several times, major reforms are needed again, underpinned by systematic prospective analyses. A novel method is needed to take into account the wide-ranging and complex factors, shaping the future of the higher education system. 'Futures' should be devised in a multi-level structure as the bulk of trends and driving forces are international in their nature and universities are embedded in broader socio-economic systems. This new approach is demonstrated here by devising 'cascading' futures for the European Union (EU), the European Research and Innovation Area (ERIA) and universities. Several advantages can be expected from this type of prospective analysis: (i) the potential changes of these broader settings, in which universities operate, as well as their impacts on higher education can be explored; (ii) the huge diversity of higher education systems and individual universities can be reflected; and (iii) the likely impacts of different policy options can also be explored. It is also proposed to select foresight programmes from the 'prospective toolkit', given their specific features and benefits compared to other prospective methods.

Introduction

Higher education has long been dominated by universities (and colleges), but different types of new players entered this arena already in the twentieth century, and this diversification is likely to continue. This development alone would be sufficient to instigate prospective analyses on the future of higher education (HE), but other factors – demographic, technological, societal trends, and new performance metrics, financial/managerial requirements *vis-à-vis* HE organisations, etc. – are also changing the landscape for universities.[1] Further, the notions of knowledge and knowledge production are being intensely discussed, given the emerging roles of new actors in producing, disseminating, using and validating knowledge. All these developments call for strategic thinking and actions both at the level of HE organisations and policy-makers. In 2007, the European Union (EU) also launched a dialogue on the future of the European Research and Innovation Area (ERIA), of which universities are undoubtedly major players.[2] That process underlines the otherwise evident point that universities operate in broader socio-economic systems, and thus it is crucial to set the scene when thinking about their future. One possibility could be to treat these systems as given. The EU itself, however, is still evolving, as a result of a number of internal and external factors. This article, therefore, devises several distinct futures[3] for the EU as a starting point. It is also assumed that the European Research and Innovation Area can evolve in different directions, depending on the main features of the EU to a significant extent, but obviously having its own dynamics, too. The scenarios developed for universities below are largely driven by these broader structures, that is, the EU and the ERIA.

The article is organised as follows: first, the need for prospective analyses is highlighted by reviewing major trends, and drivers for future changes. Then the relevance of the proposed three-level structure is discussed. It is followed by multiple visions for the EU, the ERIA, and universities, with the time horizon of 2020–2025. The differences among various prospective approaches are also considered, focussing on the specific features of foresight processes, and their benefits for building futures for the HE sector. Finally, methodological conclusions are drawn.

Three major limitations should also be indicated. First, this article is aimed at a prospective analysis, and hence it does not offer a thorough enquiry of the current state and performance of universities. Second, a geographical limitation also applies: EU universities are considered here – but the proposed method can be extended to other regions of the world, obviously with important modifications; e.g. some EU-specific trends, policies and governance modes need to be replaced by the relevant ones. Third, it is not possible to reflect the significant diversity across universities and countries. At least three major types of diversities can be mentioned: the share of HE organisations in research activities (*vis-à-vis* the other research performing sectors);[4] the distribution of legal and financial competences between national and regional governments to regulate and fund universities;[5] and the quality of education and research activities of HE organisations.[6]

The need for prospective analysis on higher education: trends and challenges

New scientific knowledge has been produced, validated and disseminated mainly by universities for centuries.[7] In the last two centuries, however, universities have lost their monopolies. First, other research actors have emerged since the late nineteenth century, notably firms and public laboratories. Non-governmental organisations (NGOs) have also entered the research arena by the end of the twentieth century. The role of users in the innovation process is also recognised now, and is becoming much better understood (Von Hippel 1988; Fagerberg, Mowery, and Nelson 2005). Second, new actors offer higher education services, especially since the second half of the twentieth century.

Of these two main fields of activities, only education is considered in this article, when devising multiple futures for universities, to reduce complexity.[8] In the same vein, solely those trends and challenges that affect education activities are discussed below.[9] The order in which they are tackled does not necessarily reflect their significance.

Key trends and likely future changes

The most important driving forces for the current and future changes can be derived by considering the increasingly intense global competition in higher education activities; as well as financial, demographic, technological and societal factors. There is an ever more intense quest for excellence in order to improve academic recognition and thus be able to raise funds. This is adding thrust to the strong pressure for intense international collaboration, and in the meantime a fierce competition for talents (both students and staff members) among universities. Governments are under pressure to cut public expenditures to balance their budgets, and/or make tax cuts possible. Tighter funding opportunities, in turn, lead to increased competition among HE organisations for restricted funds. In the meantime, technological changes offer more sophisticated, more expensive equipment and it also becomes a must to purchase these pieces of advanced equipment, given the intense international rivalry among HE actors. That is putting extra burdens on the already tight budgets of universities. Cost-efficiency of education thus becomes a major objective: the combined effects of technological and demographic changes, aggravated by the pressure on public funding, open a gap between increasing HE costs and parsimonious public budgets allocated to HE activities.

Regional, national or supra-national policies can toughen some of the above driving forces, slow down or divert their impacts, or create new drivers for change by introducing far-reaching and resolute goals for universities. A prime example of major impacts triggered by public policies is the Bologna Process (Alesi et al. 2005).

These science and technology (S&T), societal and economic factors – coupled with various policies affecting the HE sector – give rise to several already visible trends and might cause major changes in the future.

Universities perform various tasks: teaching, academic research, joint research, technological development and innovation (RTDI) projects with businesses; scientific advice for NGOs and policy-makers. The balance between these roles is changing, leading to different 'portfolios' at different types of universities, e.g. the dominance of undergraduate teaching in some cases, and research – coupled with post-graduate training – in others.

In the meantime, teaching activities themselves are likely to change: new types of courses/degrees are to be offered to meet new societal and economic needs. The most obvious examples include short(er), more practical courses for job-seekers and regular retraining of middle and top-level managers and policy-makers, as well as researchers (as required by life-long learning). Courses for self-development would be more in demand. Further, courses/curricula should be tailored to students' needs. More pronounced demand is likely to be 'attached' to HE funds coming from governments, businesses, foundations, alumni associations, and 'consumers' (students and/or their parents). As part of that trend, stronger, better articulated needs would emerge for multi- (trans-; inter-) disciplinary education and training.

An increasing share of the 18–29 year-old age group is registered for university courses. In a number of countries, there is already a very high enrolment in HE, leading to a fundamental transition from elite universities of the previous centuries to 'mass production' of degrees. With 30–48% of the relevant age cohort attending tertiary education in most countries of the Organisation for Economic Co-operation and Development (OECD), it is neither exclusively the 'elite', who participates in it, and nor is the only aim to reproduce the academic and societal elite. Teaching and research nowadays are only 'intertwined' at a fewer number of universities and usually only at the post-graduate level. The Humboldtian model has thus become an exception, rather than the rule.[10]

The overall trend of ageing population persisting in most EU and other advanced countries would increase the share of 'mature' students. Life-long learning further reinforces this trend. Thus new methods/approaches in HE are likely to emerge; just as new types of communications between HE personnel (teachers and support/administrative staff) and students, as well as among students (from different age groups, with dissimilar experience, behavioural norms, values and culture).[11]

The number of globally active HE organisations may increase significantly, leading to a more intense competition. First, a number of HE organisations located in less developed countries are likely to accumulate respect in the global HE arena and thus emerge as non-negligible competitors. Second, new HE service providers might also evolve:

- fundamentally restructured universities, e.g. financially weak, formerly independent universities taken over by strong performers from either the HE sector, or other businesses, such as publishing houses, with the ambition of selling education services;
- newly set up 'branch' campuses of highly respected universities, using their 'parent' university's curricula;[12]
- organisers of studies/degrees,[13] as well as 'accreditation' organisations granting certificates, diplomas, or even degrees,[14] all operating without their own academic staff and courses;
- large firms setting up their own 'universities';
- NGOs setting up virtual universities.

Currently 'unthinkable' players might also launch HE services by modifying the prevailing organisational forms and 'business models' or inventing new ones. That could change the HE 'ecology' quite radically, e.g. in terms of more pronounced variety, as well as new opportunities and rules for co-operation and competition. Several HE actors, or even 'species' of HE actors, might disappear, as they are becoming 'unfit' for the radically changing environment.

Depending on the speed and extent of changes envisaged above, universities are likely to be evaluated by using new metrics, besides the conventional criteria of academic excellence (publications and citations). In particular: to what extent do they fulfil their various societal roles; what types of courses are offered to whom, at what level of quality; are they attractive for foreign staff and students; to what extent are they engaged in multi- (trans-; inter-) disciplinary training; are they using resources in an efficient way?[15] Various types of universities (e.g. ones focussing on vocational training as opposed to post-graduate teaching and research; or meeting local needs vs acting as a global player, etc.) are likely to be evaluated by different sets of criteria.[16]

There is already a wide variety of governance models (different ways of involving stakeholders) as well as management models (collegial vs professional, and their different 'blends' – see Kehm (2006)). The inherent tension between the interests, values and goals of different stakeholders, as well as the one between the need to monitor and control the various activities of universities for managerial purposes and the nature of academic activities would most likely be resolved in distinct ways by different HE actors. The emergence of new players – and new business models for universities – is likely to add 'more colours' to this picture.

In sum, sweeping changes are already occurring in the HE sector and even more far-reaching and fundamental shifts can be expected. In the meantime, the broader socio-economic environment of HE organisations is also undergoing thorough and wide-ranging transformations. The very notion of education, knowledge and research is being redefined. Stakeholders are reshaping science–society links and thus new societal demands emerge for universities. Further, the principles of the so-called new public management (accountability, transparency, efficiency and effectiveness, responsiveness, as well as forward look) are posing new requirements for HE policy-makers and managers.

Universities cannot stand still amidst these sea changes; on the contrary, they need to face new realities, either simply by reacting, or taking the initiative in a pro-active way. Indeed, a strong consensus appears to be emerging on the need for major reforms from all corners: policy-makers, analysts and universities themselves (Aghion et al. 2007; European Commission (EC) 2003, 2007; OECD 2006; League of European Research Universities (LERU) 2006; STRATA-ETAN 2002; Vincent-Lancrin 2006).

These responses/reforms, in turn, should be underpinned by relevant prospective analyses and strategic processes. Given the diversity of HE organisations themselves, as well as that of their broader context, in which they operate, it would be a gross mistake to search for a unified, 'one size fits all' solution. On top of that, the methodologies applied for guiding strategy-building processes are also diverse. Thus a conscious, well-considered decision is needed when selecting methods for prospective analyses.

The relevance of multiple, multi-level futures on HE

Methodological choices

Three major choices are to be made when devising the methodology for prospective analyses. First, is it sufficient to construct a single vision of a feasible and desired future, or is it more appropriate to build multiple futures? Second, the level of analysis should be defined. One possibility is to consider a single entity[17] on its own. Another one is to emphasise the importance of various driving forces arising from different levels of the broader systems, in which a given entity is embedded in. The third choice concerns the number and range of participants: is it only a few experts who are to be involved, or is it a participatory process, in which representatives of different stakeholders groups are engaged in structured dialogues? The first two questions are considered here, while the third one is discussed in a separate section.

Before tackling these questions, it seems necessary to state that this article takes as an 'axiom' that HE reforms, or strategies of single universities – more generally, strategic actions in any field – should rest on explicit, systematic analysis of a desired future state. Yet, a number of – potentially influential – proposals or policy documents on the HE sector do not discuss the future at all (Aghion et al. 2007; EC 2003).

As for the choice between *single* vs multiple *futures*, two hypothetical 'worlds' can be thought of. In the first one, the already detected trends are well-understood and are likely to be continued, without noteworthy new driving forces emerging, and hence no major interruptions can be expected.[18] In this world, simplicity rules, and thus extrapolation seems to be a relevant analytical tool. In the other one there are multifaceted, difficult-to-predict interactions among the identified current trends, and new major trends cause discontinuities, thus adding to the already existing complexity. A single vision, therefore, cannot provide sufficient guidance to decision-makers: multiple futures should be explored.

Futures developed by prospective analysis can be direct inputs for policy preparation or strategy-building processes: once a favourable future (future state) is identified among the feasible ones, the path(s) leading to that specific future state can be designed: a series of actions can be determined, which are likely to increase the probability of achieving the desired future. Equally, futures can be used as a sort of wake up call: in case the current trends continue – because no actions are taken to change the course – we arrive at an undesirable future state.

Given the complexity of the current and likely future trends affecting the HE sector multiple futures are needed to underpin HE reforms or strategic responses of single universities. Moreover, multiple futures are crucial tools for decision-makers even in those cases, in which the underlying trends are easy to understand and their impacts can be forecast with satisfactory precision: the likely impacts of distinct options still have to be considered.[19]

Against this backdrop, it is disquieting that major policy documents on HE do not consider multiple futures; a striking example of that approach is the EU Green Paper on the European Research Area (EC 2007).

As for the *level of analysis*, the time and resources available are decisive factors. Given these constraints, one can easily understand if a single university opts for a less demanding prospective analysis, i.e. only builds multiple visions at its own level. Or in the same vein, an umbrella organisation for universities devise futures at the level of the HE sector, disregarding trends and driving forces occurring in the broader context, affecting the HE sector as a whole. A thorough review has found an impressive number of prospective projects on HE, as well as a wide variety of organisations launching these exercises, ranging from single universities to international organisations (Georghiou and Cassingena Harper 2006). Yet, in all these cases the 'unit of analysis' is either an existing or a hypothetical university. This approach disregards two important fact(or)s, and hence leads to major shortcomings. First, universities operate in broader socio-economic systems and thus it is crucial to consider the possible developments in those structures, which they are embedded in. Without that, the likely impacts of major changes in these broader systems cannot be analysed. Second, a huge diversity can be observed among the broad HE models across continents, or across countries on the same continent. Moreover, significant differences can be observed even inside countries concerning the performance, funding and governance models of their HE organisations. When an existing or an 'abstract' university is taken as a unit of analysis, this diversity simply cannot be reflected.

This article proposes a multi-level prospective analysis – using universities in the EU as an illustration – to rectify these shortcomings. The above list of factors that shape the future of universities clearly suggests that a number of trends or challenges are international in their nature, while the legal competences to set policies are with the national or (sub-national) regional governments. The European Commission has launched several initiatives to align these regional and national policies.[20] Moreover, the EU itself is still evolving; triggered by both internal and external factors. The nature of the broader EU strategies would also determine the range of stakeholders to be involved in a participatory prospective analysis: the role and influence of university staff, students and the civil society at large, policy-makers or businesses might differ significantly in distinct 'futures' for the EU. The starting point of the proposed multi-level prospective analysis is, therefore, the EU, as the broadest socio-economic context for the EU universities. It is followed by scenarios for ERIA, as the more immediate framework in which EU universities co-operate and compete with other HE actors.[21] Finally, futures are devised for universities.[22]

Benefits of multi-level futures for HE stakeholders

The benefits of building multi-level futures are also discussed in the context of the EU, but most of them would be applicable in other Triad regions, too.

For *citizens*, as well as for *decision-makers in general*, a main advantage would be that major strategic decisions – in this case on the overall rationale of the EU policies and on the mission of the European Innovation and Research Area – are taken in a transparent and conscious way. No doubt, the 'small-scale' decisions – made every day, without taking into account the 'broader picture' – would shape these broader systems, too. This 'muddling through' might seem to be preferable for those who would like to spare the time needed for social dialogues on clearly formulated multiple strategic options, and/or want to avoid the potential tensions occurring while discussing actions and their consequences. The genuine cost, however, can be a missed opportunity: conscious, well-articulated and broadly supported strategic decisions might lead to a much more favourable future state, as opposed to the outcome of 'muddling through'.

A major benefit for *policy-makers* could be to 'simulate' the likely impacts of their decisions, by changing the various 'parameters', e.g. the overall rationale of the EU or national policies (i.e. 'switching' between different EU futures), or the actual higher education policy tools, as well as the links between HE policies, *per se*, and other policies affecting education, R&D and innovation processes (e.g. exploring the impacts of certain polices on the mobility of researchers and students). As already stressed, a number of drivers are global (or EU-wide) in their character, while decision-making competences are with the national or regional authorities. Thus, multi-level governance should be understood in order to devise appropriate policies (accomplishing what is possible and not striving for unattainable goals). A set of futures, representing the various levels of governance, can contribute to design relevant policies. Further, the diversity of universities can also be taken into account, provided that the appropriate 'ideal types' of universities are identified – and used as 'input data' for this qualitative simulation – for a specific policy design task. EU policy-makers might also use multi-level futures as a tool assisting their initiatives to align national policies. In that case, however, the national level needs to be introduced.

Stakeholders of universities – their executives, academic and supporting staff, students, businesses, the relevant community around them, be it local, regional or national – would better understand the context they work in, including the potential future states, towards which these broader systems might evolve. Hence, they would be better equipped to devise a 'future-proof', robust strategy: they can explore how their planned strategy would work in different future environments, and thus can adjust their strategy to make it successful in these different futures. They can also 'hold' the environment as given, and study if changing their university's main features would improve (or deteriorate) their performance. That would be a useful exercise in case they intend to reform their university.

Finally, for *innovation policy analysts* this multi-level structure offers three advantages: (i) the likely impacts of potential changes in the broader socio-economic systems, in which universities operate, can be analysed; (ii) the observed diversity of higher education systems and/or individual universities can be reflected by identifying relevant ideal types (tailored to specific research questions); (iii) the role of other research actors, and more importantly, the links among universities and those other research players can be discussed systematically.[23]

'Cascading' futures for HE and its broader landscape

First scenarios are devised on the EU as a whole. Then the ERIA is taken into account as a 'mezzo level' system. Finally, the most important trends and drivers are addressed at the level of universities. This simple exercise has twofold aims: (i) the futures presented here are aimed at triggering a debate among experts on the relevance of this method; (ii) the content of these futures can be used as an input for actual prospective analyses.[24]

To avoid potential misunderstandings, the major underlying assumptions are spelt out before building futures in this multi-level structure. First, as already stated, policies can modulate – e.g. speed up, slow down or 're-direct' – trends, and can also trigger new developments themselves. Second, universities cannot operate fully isolated from their socio-economic environment. For these two reasons, various EU polices under the label of the Lisbon Process, especially concerning the relative weight of competitiveness[25] and cohesion objectives, as well as the more specific ones on the ERIA are considered here.[26] Third, the interrelations between competitiveness and cohesion can be thought of in different ways: (i) as mutually exclusive goals (a 'zero-sum game', in which these policies are competing for the same set of scarce political, intellectual, organisational and financial resources); or (ii) as mutually reinforcing ones (a competitive, thriving EU can afford to support cohesion regions, while narrowing the gap between advanced and laggard regions would enhance the competitiveness of the EU as a whole). The latter view is taken here, and thus a great significance is attributed to innovation processes in the cohesion regions/countries, as well as to the wide range of policies to promote innovation. Fourth, cohesion is an issue (a) inside the large, advanced EU member states (given the significant differences among their own regions), (b) for the four 'classic' cohesion countries, and (c) for the 12 new member states. Thus, it has long been a major political and policy issue for a non-negligible part of the EU15, too – and has become even more prominent since 2004. Moreover, the forthcoming enlargement(s) would add more countries and regions to this list. Fifth, it has been shown that promoting RTDI efforts in cohesion regions via joint research projects (funded for example by the EU RTD Framework Programmes) has not compromised scientific excellence (Sharp 1998). Sixth, a pronounced policy emphasis on cohesion does – and should – not preclude competition among universities.

Futures for the EU

The point of departure is two fundamental features of the EU: (1) its main strategic intention/orientation in terms of putting the main emphasis on cohesion (societal issues) *vis-à-vis* competitiveness; and (2) its overall performance compared to the other Triad regions.[27] Four fundamentally different futures can be derived by taking into account these 'variables':

Future (A) *Double success*: A carefully balanced development strategy of the EU – composed of (i) cohesion/welfare policies pursued in a flexible way, and using appropriate, refined policy tools[28] and (ii) competitiveness policies – leads to an 'externally' successful and cohesive EU.[29]

Future (B) *Successful multi-speed EU*: The already successful EU regions, perceived as 'engines of growth', are heavily promoted by EU policies, making them even stronger, leading to enhanced competitiveness of the EU *vis-à-vis* the Triad regions. In the meantime, the gap between these successful EU regions and the less developed ones significantly widens, even inside the large, advanced member states.[30]

Future (C): The EU development strategy is incapable of harmonising the requirements of competitiveness and cohesion; policies meant to support the latter are not modernised and thus take up too much resources, and hamper the processes required for an enhanced competitiveness. Two 'variations' on this theme can be thought of:

(a) *Shaky cohesion*: Temporary success in terms of stronger cohesion (at the expense of external competitiveness and thus being 'shaky').
(b) *Double failure*: Inappropriate strategies, insufficient co-ordination of policies, poor implementation and/or external factors lead to an overall failure both in terms of cohesion and performance *vis-à-vis* the other Triad regions.

Future (D) *Failed multi-speed EU*: A multi-speed EU strategy fails to improve the performance *vis-à-vis* the other Triad regions, while it widens the gap between the advanced and less developed EU regions. The reasons for this failure can be numerous, e.g. internal (inappropriate policies and/or poor implementation), external (improving EU performance, but an even quicker development of the other Triad regions). The former case is an 'absolute' failure, while the latter is a 'relative' one. Key players of strong EU regions would act together both at an intra-regional and an inter-regional level – probably also with their counterparts outside of the EU.

None of these futures can be dismissed on logical grounds. Their likelihood, however, might differ a lot, and there is no sound method to predict which of them is most likely to materialise. The actual relevance of them is to present stark choices in terms of strategies, and project the repercussions of the strategic choices made now. In that way, these scenarios can inform present-day decisions and also show the possibilities to shape our future.

Visions for the ERIA

These different futures for the EU have strong implications for the ERIA too. In principle, therefore, different types of ERIAs can be derived from them.[31] In practice, however, not all of them are equally relevant from a strategy point of view. Thus, to demonstrate the use of the proposed method, it is sufficient to consider two EU-level futures when building ERIA visions: (A) *Double success* and (B) *Successful multi-speed EU*. Some of the main features of the types of ERIA 'fitting' to these two EU futures are presented in Table 1.[32]

Futures for universities

Taking into account the trends and drivers identified above, several futures for universities can be elaborated, depending on the extent to which the diversity of universities is taken into account. A relevant method to deal with diversity is to identify ideal types. To keep the discussion relatively simple and short, only two types of universities are considered here:

(1) Universities remain largely unchanged, performing the same functions in roughly the same organisational attributes;
(2) Universities reform themselves – or are reformed by other actors – radically by transforming their main functions and/or organisational attributes.

In other words, a sort of 'average' university is assumed when discussing *unchanged universities*: neither an extremely inward-looking, inflexible one, further characterised by inertia and poor performance, and nor a flexible, dynamic, highly successful, particularly active one in various networks – although we can find such universities at the extreme. *Radically reformed universities*, by contrast, are highly flexible and thus adapt their courses, teaching and research approaches, as well as their organisational structures, managerial practices and other internal processes to the ever changing external environment, expressed by the needs of their 'clients' (that is, students, the wider research community, businesses, policy-makers and the civil society). They possess excellent 'navigation' skills to find their way in this complex world, often characterised by conflicting requirements of the various stakeholders.

In this logic, a third option – to emphasise the possibility for fundamentally different futures, and thus encourage 'outside the box' thinking – could be that universities disappear and their functions are assumed by new players, who perform their tasks/roles in radically novel and diverse ways (Havas 2006).

For an actual prospective analysis, aimed at assisting decision-making either at the level of universities, regional, national or EU (ERIA) policies, a much better refined set of ideal types should be developed. The aim of the above 'crude' typology is just to demonstrate that (a) different types of universities would act in different ways in the framework of the same ERIA; and (b) the same type of universities would behave differently – at least to some extent – when they are embedded in different socio-economic systems.[33] In other words, this method can be understood as a sort of qualitative simulation. Thus, the method itself should not be judged by the choice of these simplified types of universities, taken as somewhat arbitrary 'inputs' for this qualitative simulation.

Table 1. Features of the ERIA in two EU futures: 'Double success' vs 'Successful multi-speed EU'.

	EU	
ERIA	'Double success'	'Successful multi-speed EU'
Rationale for EU RTDI policies	'Double-track': tackle societal challenges, promote cohesion and enhance competitiveness	Excessive emphasis on enhancing competitiveness
Location of major HE and research centres	Widely distributed across the EU, weaker centres are strengthened, new ones are set up in laggard regions with a specific objective to promote cohesion	Concentrated in already strong, successful regions
Research agenda	An appropriate balance between societal and techno-economic issues	Focus on techno-economic issues; some research efforts to tackle social challenges stemming from the widening gaps between flourishing and laggard EU-regions
Mobility of researchers, university staff and students	'Two-way traffic': gaining experience, building contacts in more advanced regions across the Triad, and then exploiting these contacts upon return to 'cohesion' regions via intense, mutually beneficial co-operation Mobility grants explicitly aim at nurturing talents for excellence in RTDI and competitiveness and fostering cohesion	'One-way street': brain-drain from laggard regions to booming ones Policy schemes aim at further strengthening strong regions via mobility grants 'Two-way traffic' with strong Triad countries/regions
RTDI collaborations	Widely occurring across the EU and globally; policies aimed at promoting RTDI collaborations have an explicit aim of fostering cohesion, too, among other EU-wide issues	Mainly among strong, successful regions across the Triad, driven by businesses, supported by policies; laggards are left out
Innovation systems, co-operation among key players[a]	Strong, flexible innovation systems in a large number of regions (with their own specific strengths), capable of renewal and adaptation to the external environment, underpinning both cohesion and competitiveness Intense communication among businesses, academia, policy-makers, and the civil society to set RTDI priorities – relevant for cohesion and competitiveness–; strong academia-industry co-operation, mutually beneficial, intense links among large firms and SMEs in a large number of regions (gradually increasing over time) Co-ordinated, joint efforts – supported by EU funds – to strengthen weaker innovation systems, including communication, networking and co-operation among key players inside those regions and across regions	Strong, flexible innovation systems in the advanced regions, capable of renewal and adaptation to the external environment, underpinning sustained competitiveness Intense communication among businesses, academia, and policy-makers to set RTDI priorities relevant for enhancing competitiveness; strong academia-industry co-operation, mutually beneficial, intense links among large firms and SMEs both inside and across flourishing regions Ad hoc, weak communication and co-operation among the key players in laggard regions; weak RTDI policy constituencies Insufficient, half-hearted EU-supported efforts – at best – to strengthen weaker innovation systems of laggard regions/countries
Financial infrastructure	Conscious EU efforts (policies, guidelines, networking, exchange of experience) to improve financial infrastructure across the EU	No conscious EU efforts to improve financial infrastructure in the laggard regions
Policy-preparation methods, practices	Conscious EU efforts (guidelines, networking, exchange of experience) to improve policy-making practices across the EU	No conscious EU efforts (guidelines, networking, exchange of experience) to improve policy-making practices in the laggard regions

[a]Co-operation with the relevant Triad partners is taken for granted, i.e. not discussed here as a distinguishing feature.

Tables 2–3 identify major changes in the external environment of universities – relying on multiple visions developed for the EU and ERIA – and explore the likely features of unchanged and radically reformed universities under those conditions, focussing on their education activities.[34]

Foresight and other prospective techniques

As already pointed out futures-building (prospective) activities can be conducted by a small group of experts, or by involving the representatives of key stakeholder groups. This section briefly addresses the basic differences between these approaches, highlighting the specific features of foresight programmes/projects, as well as their expected benefits.

Locating foresight among prospective analyses

Decision-makers, experts and laymen living in different historical periods and in different socio-economic systems shared at least one desire: to know their future in advance or even to influence it for their advantage. They had used very different approaches and methods from spiritual/religious ones to scientific investigations and various modes of planning. Without going into detail, it is worth recalling some of the major methods/approaches in order to locate – and distinguish – foresight programmes:

- visionary thinking (prophets in ancient times, consultants in modern times)
- forecasting
- futures studies (for pure academic purposes)
- prospective analyses (for business or policy purposes, e.g. roadmapping, list of critical/key technologies)
- strategy formation (at firm, sectoral, regional or national levels)
- scenario planning (at a firm level)
- indicative national planning
- central planning (at a national level)
- foresight programmes.[35]

Obviously, the above approaches have a number of common characteristics. All of them (a) deal with the future(s) in one way or another; (b) collect and analyse various pieces of information, and (c) can apply a wide range of methods. Three key features can be used to differentiate the above approaches and thus distinguish foresight programmes from other methods.

These approaches can:

(1) be action-oriented vs 'contemplative' (passive)
(2) be participatory vs non-participatory
(3) consider multiple futures vs a single future state (already 'set' by external forces).

Action-oriented endeavours aim at shaping/influencing the future,[36] while passive ones are 'contemplating' about it (futures studies, without being connected to decision-preparatory processes). In other words, the latter ones merely try to develop a better-informed anticipation of the future.

Participatory future-oriented programmes/projects meet the following three criteria: they (i) involve participants from at least two different stakeholder groups in structured, face-to-face dialogues; (ii) disseminate their preliminary results (e.g. analyses and tentative conclusions) among

Table 2. Driving forces and their likely impacts on universities: 'Double success' case.

	Universities	
Trends, driving forces	Largely unchanged universities	Radically reformed universities
The role/mission of universities	The main emphasis is on teaching and 'basic research' (science for the sake of science), not much interaction with other players in (regional, national, sectoral, international) innovation systems and with the society Universities do not understand/take on their role in addressing societal issues Increasing tensions between these 'traditional' universities and the societal and techno-economic requirements of an ERIA in the *Double success* EU	A new balance of the main activities; and a new way to conduct them: intense interactions with other players in (regional, national, sectoral, international) innovation systems and with the society New activities to promote cohesion among EU regions and enhance competitiveness in the meantime Universities understand the societal and techno-economic requirements of an ERIA in the *Double success* EU, and able to adapt to this new environment
Competition for talents	Only a few 'world-class' EU universities can attract talents from advanced Triad regions Mindsets are against competition, measurement and evaluation – beyond the traditional academic indicators Inferior performance and a weakening position *vis-à-vis* the leading Triad universities	A large(r) number of EU universities become attractive for talents from advanced Triad regions Universities focussing on serving regional/local needs do not pay attention to attract talents from other countries Competition, measurement and evaluation of performance is widely accepted Strong performance *vis-à-vis* the leading Triad universities
Courses/degrees	Mainly 'traditional' courses/degrees are offered, following a 'pure science' rationale; i.e. societal needs and competitiveness issues are largely neglected Shorter, more practical courses are missing or exceptional Life-long learning is perceived as a challenge to centuries-long traditions, and not taken as a great opportunity	Teaching programmes are balanced in terms of meeting societal and techno-economic (competitiveness) objectives Life-long learning becomes a reality; most universities across the EU are flexible enough to offer the right mix of longer (traditional) and shorter courses, adjusted to the new structure/balance of learning and working
Multi-disciplinary education/training	Multi-disciplinary education slowly becomes a more widely used practice, but limited to the logic of 'pure science'. In other words, the complexities of societal issues and competitiveness are not addressed	Multi-disciplinary education becomes a widely used practice. Students are trained to understand the close relationships between societal and techno-economic issues/challenges

Table 3. Driving forces and their likely impacts on universities: 'Successful multi-speed EU'.

	Universities	
Trends, driving forces	Largely unchanged universities	Radically reformed universities
The role/ mission of universities	The main emphasis is on teaching and 'basic research', not much interaction with other players in innovation systems and with the society Some of the 'elite' universities put emphasis only on enhancing competitiveness	Emphasis on enhancing the competitiveness of businesses; all activities serve this goal; close co-operation with businesses
Competition for talents	Same as in the *Double success* case	Same as in the *Double success* case
Courses/degrees	Same as in the *Double success* case	Teaching programmes put emphasis on meeting techno-economic (competitiveness) objectives at the expense of societal challenges Life-long learning is a daily practice mainly in the advanced EU regions; in the laggard ones it is available for, and requested by, only a tiny share of citizens. Universities located in the advanced regions are flexible enough to offer the right mix of longer (traditional) and shorter courses, adjusted to the new structure/balance of learning and working. Most universities located in the laggard regions are not prepared/ flexible enough to offer these 'mixes' of courses.
Multi-disciplinary education/training	Same as in the *Double success* case	Multi-disciplinary education mainly integrates disciplines relevant for tackling techno-economic (competitiveness) issues (i.e. eclipsing societal issues).

interested 'non-participants',[37] e.g. at workshops, via the internet with free access for everyone, or in the form of printed documents, leaflets, newsletters; and (iii) seek feedback from this wider circle. Conversely, if any of these criteria is not met, that activity cannot be regarded a participatory programme.

Finally, certain prospective analyses are based on the assumption that the future is not predetermined yet; and thus it can evolve in different directions, to some extent depending on the actions of various players taken 'today'. In other words, there is a certain degree of freedom in choosing among the feasible future*s* and hence increasing the chance of arriving at the preferred future state. Clearly, there is a close link between being action-oriented and considering multiple futures.[38] Other approaches, on the contrary, can only think of a single future, already 'fixed' by certain factors and thus the task is to forecast 'the' future scientifically.[39]

In sum, foresight programmes are action-oriented, participatory and consider multiple futures.

Why foresight for building HE futures: specific advantages

The very idea behind foresight programmes is to bring together different stakeholders with their diverse sets of accumulated knowledge and experience, as well as distinct viewpoints and approaches so as to enrich the discussion and analysis. The participatory element of foresight programmes is particularly important in the case of building futures for the HE sector: given the vital role of universities in generating, transmitting, disseminating and applying knowledge, and hence their contribution to socio-economic development, major stakeholders need to be involved when strategic decisions are to be made concerning universities. Further, a foresight process aligns the participating actors around emergent agendas, resulting in a co-ordinated mobilisation of people, resources and actions. The shared vision and policy recommendations, stemming from the dialogue among participants, lead to commitment to joint actions, as well as actions by individual organisations along the lines of the shared vision. This, in turn, offers a basis for faster and more efficient implementation. In contrast, futures developed by individuals can only experiment with new methods, or spark dialogues, by offering food for thought, at best.

A foresight process might have many different outcomes. Following the usual distinction in the literature, we can think of 'process benefits' and 'products'. The first would include more intense, regular communication among the stakeholders even when the process is completed; stronger co-operation; a shared vision, leading to consensus on the actions that need to be taken; commitment to act upon the recommendations emerging from the process. The second refers to lists of priorities and proposed actions (for different stakeholders, in this case e.g. university rectors and deans, regional, national and EU policy-makers, businesses and local communities as partners of universities), inputs for strategic planning (again, at different levels). The type of intended outcomes always depends on the objectives of a foresight programme, i.e. if it is mainly a process-oriented exercise, a product-oriented one, or a mixed approach is taken. Acting upon the recommendations – e.g. strategy formation for a specific university, strategies for the higher education sector in a region, a country or the EU – is the competence and responsibility the decision-makers.

Conclusions

Universities have been around since the twelfth century in Europe and must have reinvented themselves several times, rather significantly – otherwise they would have not survived. The fact that in heated disputes some politicians and observers regard them as still largely 'medieval organisations' indicates, however, that major reforms are needed again. The reasons for that are wide ranging: globalisation of the economy and research; emergence of new HE 'service providers' and research performers; changing science–society links and societal demands towards universities; demographic changes; 'massification' of higher education; student 'consumerism'; technological development (offering new opportunities, and in the meantime putting extra burdens on the already tight budgets of universities); tensions in the national/regional budgets financing higher educations, and the concomitant requirements of the so-called new public management (accountability, transparency, efficiency and effectiveness, responsiveness, as well as forward looking); and finally the new methods, approaches and norms to organise, manage, validate, legitimate and evaluate HE and research activities.

This article has argued that given the complexity of these factors, the proposed new round of HE reforms needs to be supported by thorough and systematic prospective analyses. That is, the practice observed in some potentially highly influential analyses and recommendations by leading academics, as well as in EC policy documents where no discussion is devoted to describe a desired and feasible future state is not a satisfactory basis for any policy decision. Further, it is not sufficient to devise a single future: HE is characterised by intricate interactions among the already visible trends, coupled with driving forces potentially causing discontinuities in the future. In such a world, multiple futures are needed to assist decision-making processes. In this way not only the huge diversity of higher education systems and individual universities can be reflected, but the likely impacts of different policy options can also be considered.

Universities are embedded in broader socio-economic systems. Moreover, the bulk of trends and driving forces are international in their nature. These fundamental features cannot be taken into account when futures are devised only at the level of universities – and this is a widespread practice in recent works on the future of HE. In contrast, this article has suggested constructing multi-level futures, using the case of EU universities as an illustration, to rectify these shortcomings. With this approach both the potential changes of these broader settings, in which universities operate, as well as their impacts on higher education can be explored.

The main intention of the futures presented in this article is to demonstrate how to use this new approach and initiate lively dialogues among stakeholders. The proposed three-level structure of futures – or 'cascading' visions – offer several advantages for policy-makers, the stakeholders of universities, as well as academics interested in prospective analysis of innovation systems.

The article has also argued that there are fundamental differences between foresight programmes, on the one hand, and future-oriented academic or consultancy projects, on the other. The very idea behind participatory programmes is to bring together different stakeholders with their diverse sets of accumulated knowledge and experience, as well as distinct viewpoints and approaches so as to enrich the discussion and analysis. Further, the shared vision and policy recommendations, stemming from the dialogue among participants, offer a basis for faster and more efficient implementation. Futures developed by individuals cannot bring these advantages, but could offer new methods and/or content elements, which can trigger fruitful methodological or policy dialogues.

Acknowledgements

This article draws on a report prepared for the high-level expert group (HLEG) on *The Future of Key Research Actors in the European Research Area*, financed by DG Research, EC. Comments on earlier versions by Andrea Bonaccorsi, Elie Faroult, János Gács, Annamária Inzelt and the anonymous referees are gratefully acknowledged.

Notes

1. The term 'universities' is to be understood to cover all sorts of higher education establishments, including, for example, the German *Fachhochschulen*, the British *polytechnics* and the French *Grandes Ecoles*.
2. ERIA is understood throughout this paper as the set of all relevant actors of RTDI processes in the EU, as well as their interactions. In other words, 'ERIA policies' of the EU are just one element of ERIA, as it is composed of all other EU, national and regional policies affecting RTDI processes and performance, the activities of firms, all types of R&D units, higher education organisations, financial intermediaries, as well as a host of supporting, bridging and service organisations, and most importantly the interactions (communication, networking, collaboration and competition) among these actors.
3. 'Futures' are depicting 'images of a future world', or 'views of future states'. 'Futures', 'visions' and 'scenarios' are used as interchangeable notions here. That is, 'vision' is a neutral, value-free term in this article, and 'scenarios' are distinguished from 'fully-fledged' or 'path scenarios'. The latter ones develop detailed causal stories of how an organisation (or sector, region, country, etc.) might be transformed in a certain time period: when it is likely to arrive at cross-roads, and what direction (decision) is to be taken at those points.
4. The business enterprise sector is a dominant one in the majority of OECD countries and all the advanced ones share this feature.
5. In small and medium-sized countries it is mainly the responsibility of the national government to devise HE policies, as well as to fund HE organisations, while in bigger countries – e.g. in Germany – regional authorities perform these tasks. For a more detailed discussion, see, e.g. Sanz-Menéndez (2007).
6. A detailed analysis of the performance of EU universities is offered by Bonaccorsi and Daraio (2007).
7. Inventors have certainly advanced technologies to a very significant extent. Moreover, several major inventions have long preceded scientific theories explaining their actual operation. (The well-known examples include the steam engine, the first airplanes, semiconductors, etc.) Contrary to the widespread belief that technologies are, in essence, applied sciences, a number of scientific disciplines evolved from the puzzles why certain technologies work as they do (Nelson 2004; Rosenberg 1996, 1998).
8. It is a somewhat artificial separation, no doubt. The main purpose of this paper, however, is to demonstrate the expected benefits of multi-level futures, rather than to offer full descriptions of potential future states, covering all aspects. If this approach is accepted in actual strategy development processes, the aspects of research activities can easily be added. Both education and research aspects are considered, e.g. in Havas (2006).
9. Some of them would affect both education and research activities of HE organisations, of course.
10. The Humboldtian model of universities assumes a unity of teaching and research, based on the idea of higher education through exposure to, and immersion in, research activities (Kehm 2006).
11. Some signs can already be felt in the USA: 'Institutions designed and operated for youthful students have often been traumatised by the changing composition of the student population. This is especially true of the faculties who are ill-equipped to deal with the demands of the 3 million working adult students who not only want an education but they want it delivered in much the same way the other services they purchase are delivered: efficiently, conveniently as to time and place, courteously and with a consistent structure yielding a uniform quality. Furthermore, they want an education that, quite apart from what it may do for them as reflective beings, will improve their performance in the workplace whether it be in the professions or technical position' (Sperling 1999, 114).
12. The West Report has devised a number of business models for universities, among others, this one (GAL 1999).
13. The Western Governors University is a case in point: it has been set up as a private collaborative venture by Governors of 18 states in the USA and a number of large companies. It offers distance learning courses via its web site, alongside with 'brokered' courses and degrees (provided by 'real' education institutes), and act as a clearinghouse, too (Farbman 1999).
14. These certificates, etc. can be based on proved competences, or more conventional course work done by the 'students' elsewhere, including e-learning. Offering these 'accreditation services' can be a new role for existing universities, too.
15. As research activities are not covered in this article, the respective evaluation criteria are not discussed here, either.
16. The overall rationale of ERIA, in which EU universities operate, is also likely to affect evaluation criteria and methods.
17. The unit of analysis can be an organisation, an economic sector, a region, a country, or a group of these entities.
18. Intentions of policy-makers or those of other stakeholders to change the current settings are understood here as driving forces, potentially leading to discontinuities. In other words, 'simple worlds' are characterised by the lack of these intentions.
19. Without having considered diverse options, one cannot really speak of decision-making.

20. The most visible ones are the so-called Bologna process, the regular meetings of education ministers, as well as the other channels of the so-called 'open method of co-ordination'. The Spring European Council meetings, assessing the progress towards the Lisbon strategy, using several indicators on HE performance, can also influence national and regional HE policies. Also indirectly – and less manifestly – the various EC funded projects and expert groups on higher education can also shape these policies.
21. Non-EU universities are already operating in the EU, and given the intense international competition for students and staff, their presence is likely to be more pronounced.
22. The national – and sub-national regional – level is 'skipped', given the huge diversity of the national (regional) education systems. Skipping these levels from this illustrative case, however, does not imply that national (regional) factors can be neglected in actual prospective analyses.
23. As already noted, the research activities of universities are not discussed in this article – yet, it is worth mentioning here this latter benefit, too.
24. It is beyond the scope of this article to estimate the degree of probability of the specific futures. The modest aim is to sketch 'consistent and coherent descriptions of multiple hypothetical futures that reflect different perspectives on past, present, and future developments, which can serve as a basis for action. They are tools for thinking about the future, which will be shaped partly through deliberate strategies and actions, partly by factors beyond the control of decision-makers' (OECD 2006, 1).
25. There is no widely accepted definition of competitiveness. Economists have dissenting views even on the appropriate level of analysis: products, firms, value chains (production networks), (sub-national) regions, nations, or even larger entities. This article cannot attempt solving this problem.
26. One of the strategic guidelines of the EU cohesion policy is to improve the knowledge and innovation for growth (EC 2005). More specific areas of interventions include: improve and increase investment in RTD, facilitate innovation and promote entrepreneurship.
27. Emerging countries, e.g. China and India, might also become important competitors, but a flexible interpretation of the Triad regions can easily include any relevant countries.
28. The cases of Denmark, Finland and Sweden points to the possibility of a successful 'reformed European socio-economic model' (Aiginger 2004; Aiginger and Guger 2005).
29. This vision requires an efficient policy co-ordination, in three ways: horizontally, i.e. across policy fields; vertically, i.e. across governance levels; and along the time dimension, too, i.e. short-, medium- and long-term policies also need to be harmonised. The vision itself, however, makes no assumption if this co-ordination is achieved via heavy-handed top-down mechanisms or as concerted actions of member states and other key players, without a strong centre. This is the well-known issue of having or not a 'federal EU'. (See also two visions of the EUROPOLIS project (2001), coined 'Federal Europe', and 'Roundtable Europe', respectively.)
30. Two types of EU behaviour can lead to this future state: (1) a conscious strategic choice to use available funds and other policy tools exclusively or excessively for boosting competitiveness, and thus ignoring cohesion; (2) incapability to devise strategies and policies, and/or general inaction, inertia, inefficiency to implement policies. In a radical scenario, not to be discussed here, the loss of EU policy-making power to national, regional, and local authorities would also result in widening gaps among regions. For a largely similar scenario, called 'Swiss Europe', see EUROPOLIS (2001).
31. As already stressed, ERIA is understood throughout this paper as the set of all relevant RTDI actors, as well as their interactions. Therefore, by making a strong link between the EU strategies on the one hand, and the ERIA, on the other, does not deny the possibility that 'ERIA policies' can enjoy some level of independence from the overall strategy of the EU. Yet, it would go beyond the scope of this paper to discuss when this potential 'discrepancy' can be seen as a 'healthy, creative' tension, i.e. when ERIA policies take the lead in the 'right' direction, and pull in other policies, too; and when it is 'destructive' by hampering development and/or leading to waste of public resources.
32. A more detailed discussion can be found in Havas (2006).
33. Universities, obviously, have a certain level of autonomy in choosing their strategies.
34. Research activities of these two types of universities, using the same structure, are considered in Havas (2006), as well as the impacts of those important driving forces, which can affect universities regardless of the differences among the multiple futures devised at EU and ERIA levels.
35. The term 'foresight programme(s)' is used here as an attempt to distinguish individual (personal) foresight and 'collective' foresight programmes, i.e. the ones launched (and sponsored) by an organisation (or several ones), and conducted by a number participants. Moreover, an increasing number of articles are published by researchers working in the field of future studies, in which 'foresight' is used as a new label for their work (although still following the 'futures studies' or futurology paradigm). Thus it might be useful to tell apart that sort of 'foresight' and 'foresight programmes'.
36. *Nota bene*, the first UK Foresight Programme had chosen the following slogan: 'Shaping our future'.

37. 'Non-participants' are those persons who have not been members of panels or working groups set up by the programme, and have not been involved directly in any other way, e.g. by answering (Delphi) questionnaires.
38. Some foresight programmes, e.g. the second Swedish Technology Foresight Programme, consider alternative futures with the explicit aim of identifying key choices confronting their 'constituency', but do not intend to single out any preferred future. In other words, these programmes do not follow a normative approach. (The author is indebted for this comment to Göran Pagels-Fick.)
39. Cuhls (2003) offers an excellent, comprehensive discussion on the differences between forecasting, prediction, planning and foresight. The possibility of a single future vs multiple futures is a central element of her analysis.

References

Aghion, Ph., M. Dewatripont, C. Hoxby, A. Mas-Colell, and A. Sapir. 2007. Why reform Europe's universities? *bruegelpolicybrief*, issue 2007/04, Breugel, Brussels. http://www.bruegel.org

Aiginger, K. 2004. Copying the US or developing a new European model – policy strategies of successful European countries in the nineties. Paper presented at the UN-ECE Spring Seminar 'Competitiveness and Economic Growth in the ECE Region', Geneva, 23 February 2004.

Aiginger, K., and A. Guger. 2005. *The European socio-economic model: differences to the USA and changes over time*. Vienna: WIFO, Austrian Institute of Economic Research.

Alesi, B., S. Bürger, B.M. Kehm, and U. Teichler. 2005. *Bachelor and Master courses in selected countries compared with Germany*. Bonn: Federal Ministry of Education and Research.

Bonaccorsi, A., and A. Daraio, eds. 2007. *Universities and strategic knowledge creation: specialization and performance in Europe*. Cheltenham, UK: Edward Elgar.

Cuhls, K. 2003. From forecasting to foresight processes: new participative foresight activities in Germany. *Journal of Forecasting* 22: 93–111.

EC. 2003. The role of universities in the Europe of knowledge. Communication from the Commission, COM(2003) 58 final. 5 February 2003, Brussels.

EC. 2005. Cohesion policy in support of growth and jobs: Community strategic guidelines, 2007-2013. COM(2005) 0299, http://www.europa.eu.int/comm/regional_policy/sources/docoffic/2007/osc/index_en.htm

EC. 2007. The European Research Area: new perspectives. Green Paper. COM(2007) 161.

EUROPOLIS. 2001. The European Research Area: A new frontier for Europe? *la lettre OST* no. 22.

Fagerberg, J., D.C. Mowery, and R.R. Nelson, eds. 2005. *The Oxford handbook of innovation*. Oxford: Oxford University Press.

Farbman, M. 1999. University of the future. In *Universities in the future*, ed. M. Thorne, 69–75. London: Foresight, OST, DTI.

GAL. 1999. The WEST Report: An economic perspective, extract from 'Australian higher education in the era of mass customization'. In *Universities in the future*, ed. M. Thorne, 76–103. London: Foresight, OST, DTI.

Georghiou, L., and J. Cassingena Harper. 2006. The higher education sector and its role in research: status and impact of future-oriented technology analysis. Anchor paper for the Second International Seminar on 'Future-Oriented Technology Analysis: Impact of FTA Approaches on Policy and Decision-Making', Seville, 28–29 September 2006. http://forera.jrc.es/documents/papers/anchor/HigherEdAnchorPaper.pdf

Havas, A. 2006. Futures for universities. Paper presented at the Second International Seminar on 'Future-Oriented Technology Analysis: Impact of FTA Approaches on Policy and Decision-Making', Seville, 28–29 September 2006. http://forera.jrc.es/documents/papers/Futures%20of%20universities_paper.pdf

Kehm, B.M. 2006. Governance and strategy. Presentation at a workshop on 'Driving Forces and Challenges for the European University', Brussels, 23–24 March 2006.

LERU. 2006. Universities and innovation: the challenge for Europe. League of European Research Universities. http://www.leru.org/?cGFnZT00

Nelson, R.R. 2004. The market economy, and the scientific commons. *Research Policy* 33: 455–471.

OECD. 2006. Four futures scenarios for higher education. OECD CERI, presented at the meeting of OECD Education Ministers, Athens, 27–28 June 2006.

Rosenberg, N. 1996. Uncertainty and technological change. In *The mosaic of economic growth*, ed. R. Landau, T. Taylor and G. Wright, 334–353. Stanford, CA: Stanford University Press.

Rosenberg, N. 1998. Technological change in chemicals. In *Chemicals and long run economic growth*, ed. A. Arora and R. Landau, 193–229. New York: Wiley.

Sanz-Menéndez, L. 2007. Regional governments. In *The future of key research actors in the European research area – working papers*, EUR 22962, pp. 207-229. Luxembourg: Office for Official Publications of the European Communities.

Sharp, M. 1998. Competitiveness and cohesion – are the two compatible? *Research Policy* 27: 569–588.

Sperling, J. 1999. A business model of higher education in 2025. In *Universities in the future*, ed. M. Thorne, 104–119. London: Foresight, OST, DTI.

STRATA-ETAN expert group 2002. Higher education and research for the ERA: current trends and challenges for the near future. http://cordis.europa.eu/foresight/reports.htm

Vincent-Lancrin, S. 2006. What is changing in academic research? Trends and futures scenarios. *European Journal of Education* 41: 169–202.

Von Hippel, E. 1988. *The sources of innovation.* Oxford: Oxford University Press.

Index

Page numbers in *Italics* represent tables.
Page numbers in **Bold** represent figures.
Page numbers followed by b represent box.

N